Dr. Maggie's GRAND TOUR of the SOLAR SYSTEM

Kane Miller
A DIVISION OF EDC PUBLISHING

For my daughter, Lori

First American Edition 2020
Kane Miller, A Division of EDC Publishing

First published in Great Britain in 2019 by Buster Books
Copyright © Buster Books 2019
Written by Dr. Maggie Aderin-Pocock
Illustrated by Chelen Écija
Designed by Kim Hankinson; Edited by Frances Evans and Katy Lennon
Cover design by John Bigwood and Angie Allison; Fact-checking by Stuart Atkinson
With special thanks to Hazel Songhurst and Sophie Schrey

For information contact:
Kane Miller, A Division of EDC Publishing
P.O. Box 470663
Tulsa, OK 74147-0663
www.kanemiller.com
www.edcpub.com
www.usbornebooksandmore.com

Library of Congress Control Number: 2019940795

Printed and bound in China
1 2 3 4 5 6 7 8 9 10

ISBN: 978-1-68464-034-8

CONTENTS

INTRODUCTION

Introducing Dr. Maggie   4
What is a Solar System?   6
How the Tour Works   8
Tour Map   10

Ready for Liftoff!   12
Getting into Orbit   14
Getting into Space   16

THE TOUR

Earth   18
The International Space Station   22
Space Junk   24
The Moon   26
The Sun   32
Mercury   38
Venus   44
Missions to the Inner Planets   50
Mars   52
Asteroid Belt   60
Rocky Bodies   62
Exploring the
Outer Solar System   64

Jupiter   66
Saturn   74
Uranus   82
Neptune   86
The Kuiper Belt   90
Pluto   92
Planet Nine   96
Where Does the
Solar System End?   98
The Oort Cloud   100
The Space
Between the Stars   102
Dr. Maggie Signing Off   104

SHIP'S DATABASE

Ship's Database   106
Space Words   118
Index   120

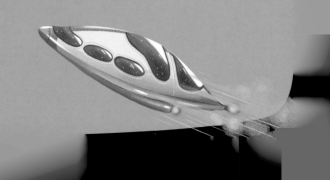

INTRODUCING
DR. MAGGIE

Hi! My name is Dr. Maggie. I am a space scientist and I have always wanted to leave the pull of Earth's gravity and travel around the Solar System.

In this book, I'm going to embark on an amazing journey – and I want YOU to come with me! We will visit planets, moons, asteroids and satellites, leaving no question unanswered and no meteorite unturned. And we will travel to places that no one has been to before. It's going to be EPIC.

Growing up, I was just an average kid, so why did I get the space bug so bad? Well, I was born during a fantastically exciting time for space exploration. The Space Race was on, the first humans had explored the Moon, and probes and satellites were traveling farther and farther into space. Ever since then, I've been looking for ways to get out there myself.

But getting into space is tricky. For one thing, only a few people have made it – just under 550 humans have traveled there. For another, space travel is incredibly expensive. The Apollo program that landed 12 men on the Moon is thought to have cost over 20 BILLION DOLLARS – and that was in the 1960s. The equivalent cost today would be around SEVEN times that amount. Not only would it take me many lifetimes to save up that much, I'd also need several more lifetimes to complete a real-life grand tour. Even with the fastest space technology, traveling at 1.5 million kilometers (km) per day, it would take me over 300 years to reach the start of the Oort Cloud at the most distant edges of the Solar System.

So how will we make this incredible journey? My inspiration came when I was reading about the scientist Albert Einstein. You know the chap – mad hair, German accent and some truly wonderful ideas. His ideas were so far out that he used "thought experiments" instead of actual experiments to test his theories. For example, to test his theory of special relativity, Einstein IMAGINED what it would be like to travel on a beam of light.

And that's exactly what we're going to do here. With imagination, we can take a journey no human or machine has made, going right to the edge of the Solar System and making it home in time for dinner.

Come on ... what are we waiting for?

WHAT IS A SOLAR SYSTEM?

Before setting off on a "Grand Tour of the Solar System," we should find out what a solar system is, and whether ours is the only one.

FOR A LONG TIME WE THOUGHT OURS MIGHT BE THE ONLY SOLAR SYSTEM IN THE UNIVERSE, BUT WITH IMPROVED TECHNOLOGY WE ARE DISCOVERING MORE AND MORE OF THEM OUT THERE.

IT STARTS WITH A STAR

We can think of a solar system as a "gravity gang" made up of a star at the center and all the planets, asteroids, dwarf planets and comets that orbit around it. It's the star's gravity that stops the other members of the gang from flying off into space.

Our Solar System is made up of the Sun (our star), the eight planets that orbit it – Mercury, Venus, Earth, Mars, Jupiter, Saturn, Uranus and Neptune – and their moons, asteroids, comets and other celestial bodies. Its "official" boundary is an area called the heliopause. This is the point where the Sun's forces give way to the influences of other stars.

EXPLORING OUR
SOLAR SYSTEM

In the 1950s, we took our first steps into space, and in the 1960s we started exploring our neighboring planets by sending probes into the inner Solar System – such as Mariner 4, below, which journeyed to Mars. In the 1970s, we sent more probes farther into space, to gain a better understanding of our place in the universe and to learn about the colder, outer planets.

A GALAXY IS A COLLECTION OF STARS AND THEIR SOLAR SYSTEMS. OUR SOLAR SYSTEM IS PART OF THE MILKY WAY GALAXY.

As we explored, we quickly realized that the universe is GINORMOUS and that our Solar System is just a very small cog in our GARGANTUAN galaxy, known as the Milky Way. Experts now know that the Milky Way is made up of billions of stars, and many of those stars have solar systems similar to our own. That means there are also billions of other planets (known as "exoplanets") in our galaxy. Some of them may even be home to alien life forms.

MOST OF THE STARS WE SEE SIT AT THE CENTER OF SOLAR SYSTEMS. A RECENT SPACE MISSION CALLED GAIA ESTIMATED THAT THERE ARE AROUND 300 BILLION STARS IN OUR MILKY WAY GALAXY ALONE.

As you can see, there's a lot to explore! Over the next few pages you'll find some more information about how our tour will work. (Even a thought experiment needs a bit of planning …)

HOW THE
TOUR WORKS

To get the most out of our trip of a lifetime, here's an overview of the incredible sights we'll take in on the tour and some handy tips to help guide us on our travels.

THE ROUTE

We're going to start the tour in familiar territory – on Earth. After a quick look around, we'll head to the Moon, our nearest neighbor, for a spot of moonwalking (all of the clips I've seen make it look like a lot of fun). From there, we'll travel toward the Sun at the heart of our Solar System. We'll learn how the Sun and the Solar System formed before we jet off to explore the remaining inner planets – Mercury, Venus and Mars.

After a quick stopover in the asteroid belt, we'll continue farther out into space to explore the planets in the outer Solar System – Jupiter, Saturn, Uranus and Neptune. Then we will venture into the strange world of the Kuiper Belt and its icy dwarf planets (including Pluto).

Tour starts here

The Sun – page 32

Saturn's rings – page 76

LOOK OUT FOR TOUR
HIGHLIGHTS THROUGHOUT
THE BOOK. THESE ARE SIGHTS
AND EXPERIENCES THAT NO SPACE
EXPLORER WOULD WANT TO
MISS. THEY'RE MARKED
WITH THIS SYMBOL:

Beyond Pluto, we'll search for the mysterious Planet Nine – in fact, it's so mysterious that we're not even sure it's out there. Then we'll cross the heliopause, considered to be the boundary of our Solar System. Only two spacecraft have ever made it this far.

Our final stop is the Oort Cloud, a vast expanse of icy objects that sits around the Solar System. Like Planet Nine, scientists think the Oort Cloud exists, but no one has ever seen it. Our discovery of it, at the edge of interstellar space, will make a spectacular ending for our tour.

The planet panel at the bottom of each destination's page will help you to keep track of where we are and how many astronomical units (or AU – see page 111) from the Sun we've traveled.

OUR SPACESHIP'S DATABASE

There is so much to do and see that I can't fit it all into the tour pages, so I have created a database at the back of the book (starting on page 106). This contains in-depth explanations and handy comparisons to help make sense of the mind-blowing facts and figures we'll encounter on our travels. If you get lost at any point of the journey, don't forget to consult it – it will help to get you back on course.

TOUR MAP

ASTEROID BELT

URANUS

SATURN

MARS

VENUS

THE SUN

MERCURY

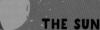

Outer planets

COUNTDOWN: 2, 1...

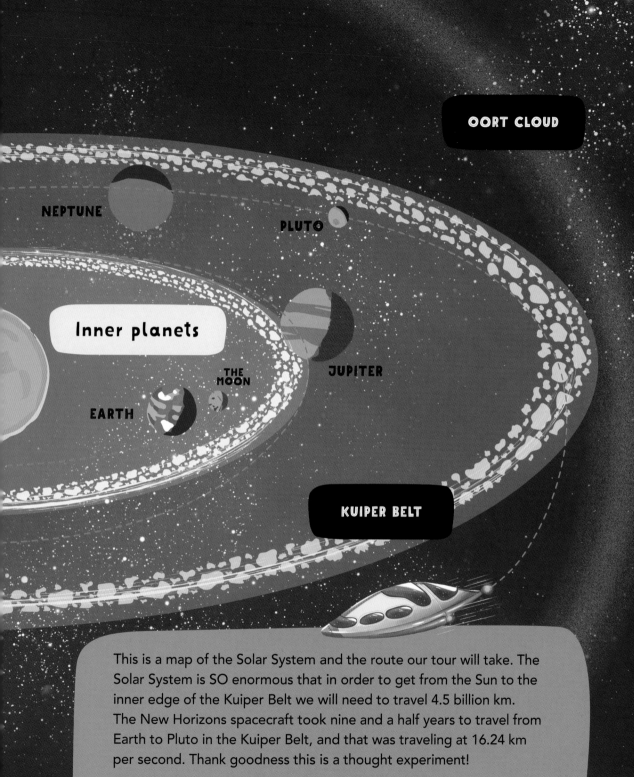

NEPTUNE

PLUTO

Inner planets

THE MOON

JUPITER

EARTH

KUIPER BELT

This is a map of the Solar System and the route our tour will take. The Solar System is SO enormous that in order to get from the Sun to the inner edge of the Kuiper Belt we will need to travel 4.5 billion km. The New Horizons spacecraft took nine and a half years to travel from Earth to Pluto in the Kuiper Belt, and that was traveling at 16.24 km per second. Thank goodness this is a thought experiment!

There's a lot to explore so we'd better get started. Grab your spacesuit and get ready for LIFTOFF!

READY FOR LIFTOFF!

While our engines warm up, let's consider how we're going to travel out into space from Earth. To do this, we need to overcome a powerful, invisible force – gravity.

THE GROUNDING FORCE

Gravity keeps us grounded – literally. It stops everyone and everything on Earth from floating off into space. The pull of gravity depends on the mass of the object it affects. Earth is GIGANTIC compared to a person, so we feel its gravitational force quite strongly. We have mass too, but we're tiny in comparison to Earth, so the force that we exert on the Earth (and on each other) is very small.

BECAUSE EARTH'S GRAVITATIONAL PULL ON US IS SO STRONG, WE NEED TO USE A LOT OF ENERGY TO TRAVEL AWAY FROM THE PLANET AND INTO SPACE.

TO LEAVE OR NOT TO LEAVE?

To start our epic journey and travel into space, we need to escape Earth's gravity. Not all spacecraft are designed to do this. Some objects are launched into space for a very short time before they fall back down to Earth – these are known as "suborbital" objects. Other objects are launched with enough power to get them high and fast enough to stay in motion (or "orbit") around the Earth. These are known as "orbital" objects.

AN OBJECT TRAVELING AT OVER 40,000 KM PER HOUR WILL LEAVE EARTH'S GRAVITY.

Orbital object

AT 20,000 KM PER HOUR, AN OBJECT WILL GO INTO ORBIT.

ANYTHING TRAVELING AT LESS THAN 20,000 KM PER HOUR WILL FALL BACK TO EARTH.

Suborbital object

GETTING INTO ORBIT

Launching an object into orbit is a bit of a balancing act. It takes an exact amount of a pushing force called "thrust" to overcome the force of gravity but, at the same time, keep an object from flying all the way out into space. Understanding how this technical challenge works will help to make sense of how the artificial satellites we'll encounter on the tour stay in orbit. Another thought experiment will help …

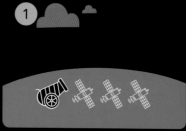

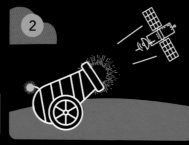

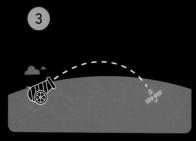

1 Let's imagine we're standing in a large, open field. A cannon and three satellites sit beside us.

2 We load a satellite into the cannon, add gunpowder and fire it into the air – BOOM!

3 The satellite zooms WAY up into the sky – then gravity pulls it back to the ground.

A SATELLITE IS AN OBJECT THAT ORBITS AROUND ANOTHER OBJECT. THE MOON IS A NATURAL SATELLITE BECAUSE IT ORBITS EARTH. THE INTERNATIONAL SPACE STATION (ISS), WHICH ALSO ORBITS EARTH, IS AN EXAMPLE OF AN ARTIFICIAL SATELLITE.

A YEAR IS DEFINED AS THE LENGTH OF TIME IT TAKES A PLANET TO ORBIT THE SUN ONCE. A DAY IS EQUAL TO THE LENGTH OF TIME A PLANET TAKES TO ROTATE ONCE ON ITS AXIS.

COUNTDOWN: Fasten your seat belt!

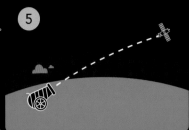

We load up the next satellite. This time we use a lot more gunpowder. We fire and … WOW! Where did it go?

Oh no, we overdid it. Our satellite overcame the force of gravity, but left Earth so fast that it zoomed beyond Earth's orbit and into space.

Let's give it another try, using less gunpowder than the last time, but more than the first time. BOOM!

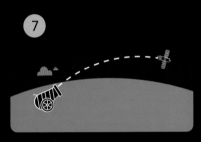

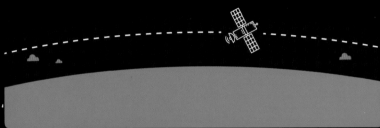

This time, the satellite is traveling fast enough to stop it from falling to the ground, but slowly enough to stop it from flying straight into space. Phew! Our satellite is now exactly where we want it – traveling around Earth in "Low Earth Orbit," about 300 km above sea level.

FREE-FALLING

Astronauts inside an orbiting spacecraft, such as the International Space Station (or ISS, see pages 22-23), float around in all directions. They are experiencing a condition known as "microgravity." They look as if they are weightless, but they're actually falling. Earth's gravity pulls them down toward it. However, the ISS, and the astronauts inside it, are falling at just the right speed to fall around the curve of Earth, rather than heading straight toward the ground. This means the astronauts are in constant free fall.

GETTING INTO SPACE

Unlike a satellite, our craft is going to escape Earth's gravitational pull altogether and blast off into space. To do this, we are going to need an ENORMOUS amount of thrust. Hold on to your helmet!

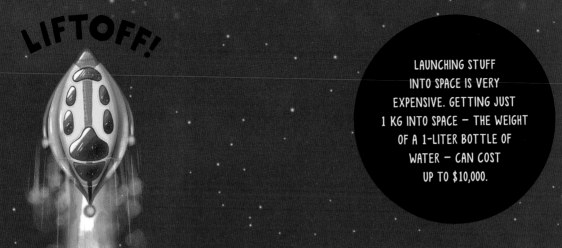

LIFTOFF!

LAUNCHING STUFF
INTO SPACE IS VERY
EXPENSIVE. GETTING JUST
1 KG INTO SPACE — THE WEIGHT
OF A 1-LITER BOTTLE OF
WATER — CAN COST
UP TO $10,000.

This is a spaceship I have designed especially for our tour. I wanted to create something that looked different than every other space vehicle out there. You can design your own craft for the trip, if you like.

Because we are using the power of thought, we don't really need a spaceship to make our journey. But we might as well travel in comfort and look the part!

EARTH

EARTH STATS

SIZE (DIAMETER)
12,756 km

DISTANCE FROM EARTH
0 km

HOW MANY BIRTHDAYS?
One birthday every 365 days

HOW LONG IS A DAY?
24 hours

1 AU FROM THE SUN

OUR HOME PLANET

And we're OFF! As we travel away from Earth's surface, let's look back for a moment at the amazing planet we call home. From up here, it looks brilliant – like a beautiful, blue globe hanging in the darkness – and a far more peaceful place than the often busy, noisy world we've left behind.

Earth is one of the solid, rocky inner planets of the Solar System. Although it feels familiar to us, it is incredibly special – we are not going to encounter another place like it on our travels.

As far as we know, Earth is the ONLY planet in the Solar System that supports life. Take another look at Earth and notice the features that allow life to thrive there: the vast quantities of water covering its surface, the thick, oxygen-rich atmosphere surrounding it, and its ideal location – neither too close nor too far from the Sun.

A UNIQUE WORLD

Before we set our course for the International Space Station, let's think a bit more about the things that make Earth so special. Understanding these features will help us to make sense of the extreme conditions we'll encounter in other places on the tour.

THE SUN

The Sun provides Earth with all the energy needed for life. It is a very reliable star and a steady source of heat and light. Earth is positioned in an area of the Solar System known as the "Goldilocks Zone," meaning it is the perfect distance from the Sun – Earth's temperature is neither too hot nor too cold, but just right for liquid water to exist.

NEIGHBORING PLANETS ALSO PROTECT EARTH. JUPITER, FOR EXAMPLE, HAS A VERY STRONG GRAVITATIONAL PULL BECAUSE IT IS SO LARGE. THIS ATTRACTS SOME OF THE ASTEROIDS AND COMETS THAT WOULD OTHERWISE HIT EARTH.

ATMOSPHERE

Our planet is surrounded by a 500-km-thick coating of gases – mainly made up of nitrogen and oxygen. This allows us to breathe, shelters us from the Sun's radiation and protects Earth from all but the largest meteorites.

THE MOON

As Earth moves around the Sun it also rotates on its axis. However, it has a slight wobble. The close proximity of the Moon stops Earth from wobbling too much. Without it, the tilt of Earth's axis would vary and the planet would have a very unstable climate.

WATER

Earth's temperature is crucial because it means water can exist on its surface as a liquid. We think liquid water is essential for all forms of life.

INSIDE EARTH

This rocky planet is made of four main layers: the outer crust, the mantle, the liquid outer core and the solid inner core. The crust is essential for living things – you and I live on it! Currents in Earth's core also create a magnetic field that protects the planet from the fierce solar wind (see pages 98-99).

THE INTERNATIONAL SPACE STATION

As we fly past the International Space Station (ISS), wave at the astronauts living on board. The ISS has been orbiting Earth since the year 2000. It is used for conducting experiments and collecting data about Earth and space.

THE ISS ORBITS THE EARTH 16 TIMES EVERY 24 HOURS. THIS MEANS THAT THE ASTRONAUTS EXPERIENCE A SUNRISE AND SUNSET EVERY 45 MINUTES. YOU CAN SEE THE ISS IN THE NIGHT SKY FROM EARTH – IT LOOKS LIKE A VERY FAST-MOVING STAR.

SCIENCE IN SPACE

Scientists on board the ISS conduct a wide range of experiments, such as testing specialist footwear, robotics and how well plants grow in space. The effects of living in space are closely monitored in the hope of finding ways for humans to live safely there full-time.

ASTRONAUTS CAN BECOME UP TO 3% TALLER IN SPACE BECAUSE THE FORCE OF GRAVITY IS NO LONGER PULLING DOWN ON THEIR BODIES. AS THEY ARE IN FREE FALL, THEIR SPINES CAN EXPAND, MAKING THEM LONGER.

DINNERTIME

There are no fridges or freezers on the ISS, so most of the food is dehydrated. This makes it lightweight and long lasting. Water is added to the food before it is eaten. Salt and pepper comes in the form of a liquid to avoid granules flying around, getting in the astronauts' eyes and clogging air vents.

LIQUID RECYCLING

Astronauts have to use a special space toilet that sucks their waste away so that it doesn't float around – YUCK! Their pee is collected and passed through a machine that recycles it into drinking water.

AS MANY AS TEN PEOPLE CAN LIVE ON THE ISS.

ARRIVAL POD

Astronauts usually stay on the ISS for six months at a time and they use a spacecraft, such as Soyuz, to get there and home again. Soyuz can carry three people plus supplies of food and water. There is always a Soyuz capsule attached to the ISS. It acts like a lifeboat to carry people to safety in case of an emergency.

SPACE JUNK

We might see some strange objects floating past us as we head toward the Moon. Space is littered with broken satellites, parts that have detached from rockets and tools that have been left by astronauts. These objects are called space junk and they can cause lots of problems. Here are some of the weirdest objects that have been found in orbit around Earth.

SCIENTISTS USE TELESCOPES FROM EARTH THAT CAN TRACK ANYTHING IN SPACE THAT IS LARGER THAN 5 CM. SO FAR, WE KNOW THERE ARE AROUND HALF A MILLION BITS AND PIECES UP IN SPACE.

SPATULA

In 2006, astronaut Piers Sellers was using a spatula to spread some protective goo onto the space shuttle. He accidentally let the spatula go and it drifted away into space.

Spatula

Camera

CAMERA

Astronaut Suni Williams was working outside the ISS in 2007 when her camera came away from her spacesuit. She wasn't able to grab it before it floated into space.

COLLISIONS

The growing amount of space debris is a big problem, especially as pieces can travel at speeds of up to 23,300 km per hour. Because there is so much stuff up there, collisions with active spacecraft and satellites are becoming more likely. Even very tiny fragments can cause a lot of damage due to the fast rate at which they are traveling.

SPACESHIP PARTS

In July 2007, astronauts on the ISS threw out a full tank of ammonia that was taking up too much room. The tank burned up when it hit the Earth's atmosphere. It must have looked like a shooting star.

Pee crystals

ASTRONAUT PEE

Before a waste-recycling system was introduced on the ISS, astronauts' pee was simply dumped overboard. It instantly froze into tiny, floating crystals.

YOU CAN SEE THE MOON BECAUSE IT REFLECTS LIGHT FROM THE SUN.

1 AU FROM THE SUN

THE MOON

EARTH'S LOCAL COMPANION

Our first destination is up ahead – Earth's closest neighbor, the Moon. The differences between this rocky satellite and our home planet are plain to see. With barely any atmosphere and not a trace of liquid water, the surface looks dry, barren and lifeless.

As we circle the Moon, we notice that – as with planets – one half is in darkness and the other is bathed in sunlight. Now we just have to decide whether to land on the day side or the night side.

MOON STATS

SIZE (DIAMETER)
3,475 km

AVERAGE DISTANCE FROM EARTH
384,000 km

HOW MANY BIRTHDAYS?
One – it follows Earth around the Sun

HOW LONG IS A DAY?
27 Earth days

THE TIDES IN THE SEAS AND OCEANS ON EARTH ARE MAINLY CAUSED BY THE PULL OF THE MOON'S GRAVITY.

MOONWALKING

We have set down our spacecraft on the cusp of where the night side meets the day side – that way we can explore the extreme temperatures in both regions. A full day on the Moon lasts for almost one Earth month, so we can take our time. Once we've got our spacesuits on it's time for EVA, which stands for "Extravehicular Activity" – let's go moonwalking!

SUN'S RAYS

The day side of the Moon is caused by the Sun's light reaching the Moon's surface. There's no protective atmosphere here like there is on Earth, so you will need to put your visor down when we enter the day side. This will help to protect your eyes from the direct glare of the Sun.

ON THE DAY SIDE ...

With no atmosphere to interrupt it, the Sun shines directly onto the surface of the Moon. It feels incredibly hot – the temperature is about 100°C, the same as boiling water.

FROZEN FOOTPRINT

There are no winds blowing over the surface of the Moon, so it is eerily still. No wind means that there is nothing, apart from collisions with asteroids and meteorites, to remove marks on the surface. This footprint, for example, was left by Buzz Aldrin in 1969 and is still there today.

YOU WILL NOTICE MANY LARGE HOLES COVERING THE MOON'S SURFACE. THESE ARE CALLED CRATERS AND ARE MADE WHEN ASTEROIDS AND METEORITES HIT THE MOON.

SPACE WALK

Walking on the Moon is very different than walking on Earth because gravity is much weaker here. The force that could send you half a meter into the air on Earth for one second, would let you leap three meters on the Moon, and the jump would last for four seconds.

THE MOON IS A GREAT PLACE TO COLLECT METEORITES. THEY STAY EXPOSED ON THE SURFACE BECAUSE THERE IS NOTHING TO COVER THEM.

ON THE NIGHT SIDE ...

In the shadows it is wonderfully cool — maybe even TOO cool! Here, temperatures can drop as low as -173°C, colder than the lowest temperatures ever recorded in Antarctica, the coldest place on Earth.

PHASES OF THE
MOON

I've spent years studying the Moon and still find it a fascinating place. It is a familiar presence in our sky and all you have to do to see it is look up! Sometimes it appears round and at other times you can only see a sliver of it. These different shapes are called the phases of the Moon.

DOES THE MOON CHANGE SHAPE?

The Moon itself doesn't change, it just looks different at different times of the month. The phase, or shape, of the Moon that we see from Earth depends on its position in relation to the Sun and Earth. Half of the Moon's surface is always lit by the Sun, but because of the way that the Moon orbits Earth, sometimes we can only see part of that lit side.

WAXING AND WANING

These words describe the different shapes of the Moon that we see in the sky. Waxing means growing and waning means shrinking.

Gibbous means swollen. In this phase, we can see nearly all of the Moon's illuminated face.

This is when half of the lit side of the Moon is visible. It is also called a half-moon.

The Moon is starting to become visible in the sky. The crescent is getting larger.

This is when the lit side of the Moon points away from the Earth.

First Quarter

Waxing Gibbous

Waxing Crescent

Full Moon

New Moon

PHASES

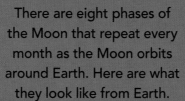

There are eight phases of the Moon that repeat every month as the Moon orbits around Earth. Here are what they look like from Earth.

Waning Gibbous

Waning Crescent

Last Quarter

This is when we can see the entire lit side of the Moon.

From this phase until the New Moon, the lit side appears to shrink.

Here, we can see exactly half of the Moon's lit side.

The Moon appears as a small crescent before the lit side disappears completely.

THE SUN

THE SOLAR SYSTEM'S HEART

From the relative comfort of the Moon, we're now traveling toward the Sun at the very center of the Solar System. The gigantic, glowing ball of gases ahead of us is a SERIOUSLY dangerous place. It acts a bit like a humungous candle, allowing us to see other bodies in the Solar System that don't give off light, such as moons and planets.

The Sun is incredibly hot. The temperature of the outer layer is 5,000°C – that is six times hotter than molten volcanic lava on Earth – and the core is a mind-boggling 15,000,000°C. The Sun constantly pours out many kinds of radiation waves, including ultraviolet rays, gamma rays and infrared light. Several of these waves are harmful to humans, but thankfully they won't penetrate our craft. Now, set the dazzle factor on your helmet's visor to maximum and see how many solar flares – gigantic flashes of light – you can count leaping from the Sun's surface.

SUN STATS

SIZE (DIAMETER)
1.39 million km

AVERAGE DISTANCE
FROM EARTH
149.6 million km

HOW MANY BIRTHDAYS?
None – the Sun doesn't have years because it is the center of the Solar System

HOW LONG IS A DAY?
24 Earth days

THE JOURNEY FROM THE MOON TO THE SUN IS AROUND 150 MILLION KM. BY PLANE, THE FLIGHT WOULD TAKE 20 YEARS.

A "CORONAL MASS EJECTION" (CME) IS SIMILAR TO A SOLAR FLARE. DURING A CME, CHARGED PARTICLES ARE RELEASED FROM THE SUN INTO SPACE. THEY CAN BE LARGER THAN THE EARTH AND SOMETIMES STOP OUR SATELLITES FROM WORKING.

AROUND ONE MILLION EARTH SPHERES COULD FIT INSIDE THE SUN. YES, IT IS THAT BIG!

A CLOSER LOOK

The Sun is a star – a burning ball of helium and hydrogen gases – so we can't land on it. But let's have a look at what we can see from a safe distance and consider what makes the Sun unique. It is the heaviest body in the Solar System; nearly 99% of all mass in the Solar System sits in the Sun. Its powerful gravity keeps the planets and other objects in orbit around it.

SUNSPOTS

Sunspots are the dark areas that are sometimes visible on the Sun's surface. They are cooler than the orange regions around them and are the focus of dramatic magnetic activity. Solar flares and violent storms erupt from sunspots.

BECAUSE THE SUN ISN'T A SOLID BODY, DIFFERENT PARTS OF IT ROTATE AT DIFFERENT SPEEDS. A SINGLE SPIN LASTS FOR 25 EARTH DAYS AT THE EQUATOR BUT 38 EARTH DAYS AT THE POLES.

OUR LOCAL STAR

The Sun is the only star in our Solar System and it is very similar to the millions of stars you can see in the night sky. The reason that the Sun looks so different than other stars when viewed from Earth is because we are so much closer to it than we are to them.

MANY OTHER SOLAR SYSTEMS IN THE UNIVERSE ORBIT AROUND MORE THAN ONE SUN. SOME DISTANT PLANETS HAVE TWO OR THREE SUNS.

G-type star

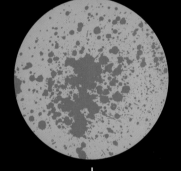
O-type star

YELLOW DWARF

Our Sun is a particular kind of star, known as a "yellow dwarf" or "G-type" star. These are some of the brightest stars around, outshining 90% of the stars in the Milky Way. Believe it or not, a G-type star is cool compared to other, even brighter types. The brightest types of star – O-type stars – have a mass 55 TIMES greater and are FIVE TIMES hotter than the Sun.

A STAR IS BORN

So, where did the Sun come from? The Sun and the rest of the Solar System formed around 4.5 BILLION years ago. Here's how it happened …

1. DUST & GAS

The Sun was formed from a GIGANTIC cloud of dust and gas called a "stellar nursery" or nebula.

2. GRAVITY

The gases inside the nebula became disturbed and began to clump together. Gravity then pulled the clumps toward one another.

THE LEFTOVER MATTER SPINNING AROUND THE NEW STAR CLUSTERED TOGETHER AND FORMED THE PLANETS, COMETS AND ASTEROIDS IN THE SOLAR SYSTEM.

3. MORE MATTER

The mass at the center attracted more matter toward it. It grew, and the pressures and temperatures inside it increased until the particles started to fuse together.

The Cat's Eye Nebula

NOT ALL NEBULAE ARE STAR NURSERIES. SOME, SUCH AS THE CAT'S EYE NEBULA, ARE KNOWN AS PLANETARY NEBULAE AND FORM WHEN A DYING STAR EXPANDS.

6. FUSION

Gravity pulled more matter to the protostar, and pressures and temperatures rose high enough to cause a process called fusion. During fusion, atoms are squished together to create new matter and release huge amounts of energy in the form of light, heat and radiation. Albert Einstein summarized this process with the famous equation: $E = mc^2$. The material was pulled toward the center of the protostar to form the new star, and the Sun was born.

5. PROTOSTAR

In the middle of the spinning disk, a protostar – a baby star – started to form.

4. SPINNING

This caused the cloud of matter to start spinning REALLY fast. This flattened it into a disk shape.

VISOR SETTING

A COLOR-ENHANCEMENT FILTER HIGHLIGHTS FEATURES ON THE SURFACE. THE CRATERS SHOW IN A BEAUTIFUL BLUE COLOR.

MERCURY

MERCURY HAS A VERY SHORT ORBIT COMPARED TO EARTH. IT ONLY TAKES 88 EARTH DAYS TO ORBIT AROUND THE SUN.

0.4 AU FROM THE SUN

MERCURY STATS

SIZE (DIAMETER)
4,879 km

AVERAGE DISTANCE FROM EARTH
91 million km

HOW MANY BIRTHDAYS?
One every three months

HOW LONG IS A DAY?
59 Earth days

A MINI PLANET

Our next stop is the smallest planet in the Solar System and the one that is closest to the Sun. Does the crater-covered surface remind you of anywhere? Don't worry, we haven't taken a wrong turn and ended up back at the Moon – we're heading toward Mercury.

Even though Mercury looks like a still, quiet place, it harbors many hidden secrets. Unfortunately, staying all day isn't an option because it would take up too much time – 59 Earth days to be exact. Let's find somewhere to land and see what we can find.

ALTHOUGH MERCURY HAS A SHORT ORBIT, IT TAKES A VERY LONG TIME FOR THE PLANET TO ROTATE ONCE ON ITS AXIS. THIS MEANS A DAY ON MERCURY IS ALMOST AS LONG AS A YEAR ON MERCURY!

MISSION ON MERCURY

All is still and quiet as we search for a good landing place. Mercury has an extremely thin atmosphere, called an "exosphere." This offers almost no protection against meteor impact, so Mercury is covered in craters. The craters will remain on Mercury's surface for billions of years. This is because, as we discovered on the Moon, no atmosphere means no weather to wear them away – so they remain untouched, unless they are hit by another meteor.

PERILOUS PITS

Watch your step as you cross Mercury's surface – it's covered in trip hazards and dangerous pits. The planet is striped with cliffs that scientists have named "lobate scarps." These scarps can be up to 3 kilometers high and hundreds of kilometers long. They were formed as the iron core of the planet cooled and contracted, causing the surface to crumple.

MERCURY, VENUS, EARTH AND MARS ARE THE INNER PLANETS IN THE SOLAR SYSTEM. THEY ARE CALLED "TERRESTRIAL PLANETS" BECAUSE OF THEIR SIMILAR HARD, ROCKY SURFACES AND METAL CORES.

STARGAZING

Look up! The stars look really brilliant here – brighter than they look from Earth. Large, hot bubbles of gas in Earth's atmosphere distort the images of stars, making them twinkle. But here on Mercury, no atmosphere means no twinkle. At certain times we can also see glowing, cloud-covered Venus and Earth farther in the distance.

THE MESSENGER PROBE FLEW PAST MERCURY TWICE BEFORE SETTLING INTO ORBIT IN 2011. IT MAPPED THE ENTIRE SURFACE BEFORE CRASHING INTO THE PLANET IN 2015 WHEN IT RAN OUT OF FUEL.

HEAVY METAL

Mercury is very small but VERY dense. It is thought to be so dense due to the presence of a large, partly liquid, iron core at its center. We can test this by taking out a compass. The needle of the compass should move because of the magnetic field produced by the planet's liquid iron core.

A STRANGE PLANET

Compared to other planets, Mercury is a bit of a peculiar place. It is a planet that has remained mysterious for some time, and only two spacecraft have visited it so far. As we take a look around, let's explore some of the strangest features of this unusual planet.

TEMPERATURE EXTREMES

The temperature differences between the day and the night sides are MASSIVE. The side facing the Sun can rocket to 450°C. However, the parts that are not bathed in sunlight can plummet to -180°C. Brrr!

ANCIENT ICE

I have a surprise for you ... as we venture into one of the craters on Mercury, you might be lucky enough to find a strange white substance. You can pick it up. It's cold and wet, a bit like a snowball, and that is EXACTLY what it is – water ice. It seems mad that we can find ice this close to the Sun, but the bottoms of these craters never see sunlight, so ice can sit here for millions of years.

EXTRAORDINARY ORBIT

Like all the planets in the Solar System, Mercury orbits the Sun in a squashed-circle shape called an ellipse. However, Mercury's orbit is unusual because it takes the planet very close to, and then very far away from, the Sun. It also travels at a superfast speed, almost 47 km per second, which is quicker than any of the other planets.

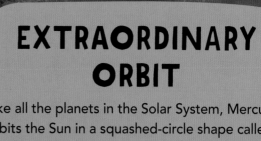

BEPICOLOMBO

To help us to learn more about mysterious Mercury, a spacecraft called BepiColombo has been launched. It is set to arrive at the planet in 2025 to begin explorations.

VENUS' SURFACE AREA ALMOST MATCHES EARTH'S. IF VENUS COULD BE PEELED LIKE AN ORANGE, ITS PEEL WOULD ALMOST COVER EARTH, APART FROM A TINY SLIVER.

VISOR SETTING

A RADAR VIEW LETS US SEE THROUGH THE CLOUDS AND STUDY DETAILS ON THE SURFACE.

VENUS STATS

SIZE (DIAMETER)
12,104 km

AVERAGE DISTANCE FROM EARTH
41 million km

HOW MANY BIRTHDAYS?
For every two birthdays on Earth, you would have three on Venus

HOW LONG IS A DAY?
243 Earth days

0.7 AU FROM THE SUN

VENUS

THE ACID PLANET

Be warned – our next destination is a classified DANGER ZONE. The first features we notice are layers of thick, yellowish clouds hiding Venus' surface. The dense cloud blanket isn't made of water vapor, as clouds are on Earth, but of sulfuric acid. This is dangerous stuff that can dissolve plastic and metal. Don't panic – because this is a thought experiment, we are quite safe and traveling to the surface won't be a problem.

As we get closer, we can use our visors to see the fierce weather systems brewing below us. These are like the hurricanes and cyclones that happen on Earth. Don't worry – things will calm down once we've made it through the upper atmosphere. Meanwhile, hold on tight!

VENUS ROTATES IN THE OPPOSITE DIRECTION TO EVERY OTHER PLANET IN THE SOLAR SYSTEM. WHY IT DOES THIS IS A MYSTERY.

WHAT'S VENUS LIKE?

Falling toward the planet's surface feels like sinking into an ocean. Our spacecraft is hit by much higher pressure than on Earth and buffeted by tornado-like winds blowing up to 300 km per hour. As we pass through the upper atmosphere, things calm down a bit … sort of. But we're not out of danger yet. Landing safely will be a challenge.

TOUCH DOWN

Although over 40 unmanned spacecraft have visited Venus, only ten have landed on the surface. The first to land (in fact, the first human-made object to touch any other planet) was Russia's Venera 3 in 1966. It didn't last long. The superhot temperatures, nasty atmosphere and a crash-landing meant that no data was transmitted back to Earth. The first craft to land successfully was Venera 7 in 1970, which transmitted data for 23 minutes before it stopped operating. It's important we keep our visit short!

VENUSIAN VIEW

The view is a bit disappointing. The sky has an orange-yellow glow, but it's impossible to see the Sun or stars through the cloud-filled atmosphere. The Venusian atmosphere is NOTHING like Earth's. It is much thicker and mainly made up of carbon dioxide and a tiny bit of nitrogen. Although carbon dioxide is found in Earth's atmosphere too, it is present in much higher levels on Venus, which would be deadly to humans.

BECAUSE VENUS ROTATES BACKWARD, IF YOU COULD SEE THE SUN THROUGH THE THICK CLOUDS (AND STAY HERE FOR A WHOLE DAY), YOU'D NOTICE IT RISES IN THE WEST AND SETS IN THE EAST.

VOLCANIC ACTIVITY

Look hard through the heat haze and you can make out ranges of volcanoes. In fact, Venus has more volcanoes than any other place in the Solar System – the tallest Venusian volcano is Maat Mons, which is almost as high as Mount Everest. There is evidence to suggest that some of these volcanoes are still active, so beware!

THE
HOTTEST
PLANET

Apart from the Sun, Venus is the hottest place we will visit in the Solar System. The average temperature here is around 450°C. When you have a pizza at home you probably set your oven to about 200°C – so Venus is more than twice as hot (cooking here would be a disaster!).

HOTTER THAN MERCURY

When we consider Venus' position in the inner Solar System, we realize that the planet is surprisingly toasty. It is hotter than the average temperature on the planet Mercury, even though Mercury is closer to the Sun.

VENUS IS THE THIRD-BRIGHTEST OBJECT IN THE SKY AFTER THE MOON AND THE SUN.

WHY IS VENUS SO HOT?

To understand why Venus is hotter than Mercury, we need to look closely at the atmospheres of the two planets. When we were on Mercury, we noticed that it resembled the Moon in many ways – not least because it had virtually no atmosphere. Venus, in contrast, has a very thick atmosphere, which is mainly made up of carbon dioxide (CO_2).

GLOBAL WARMING

Carbon dioxide is the key to Venus' superhot environment. It is also found in the atmosphere on Earth. CO_2 traps energy from the Sun in Earth's atmosphere, causing the planet's surface temperature to rise.

SEEN FROM EARTH, VENUS HAS PHASES SIMILAR TO THE MOON.

A GREENHOUSE PLANET

CO_2 molecule

CO_2 is contributing to major climate changes on Earth, but there it accounts for less than 1% of the gases in the atmosphere. Compare this to Venus, where it makes up a whopping 95%. So Venus is also suffering from an extreme form of global warming, and it is this that makes the temperature on the planet much higher than you'd expect.

MISSIONS
TO THE INNER PLANETS

For centuries, humans had to rely on their eyes or telescopes to see what was out in space. But in the 1960s, the first space probes started exploring planets up close. Some of the earliest focused on Venus and Mars.

VENUS

In 1962, the American probe Mariner 2 made a flyby trip to Venus. It was the first spacecraft to successfully fly past another planet and send back data.

Mariner 2

THIS FIRST CLOSE-UP OF VENUS WAS TAKEN BY MARINER 10 IN 1974 USING AN ULTRAVIOLET CAMERA.

Between 1961 and 1984, Russia sent 16 unmanned probes to Venus, as part of its Venera program. Ten of the probes landed successfully.

NASA and ROSCOSMOS (the Russian Space Agency) are currently discussing a new joint mission to Venus, which would launch in the 2020s.

MARS

In 1965, the probe Mariner 4 took photos of the Martian surface – they were the first close-up views of another planet. In the 1970s, two spacecraft named Viking 1 and Viking 2 landed on Mars. Part of their mission was to test the soil to look for signs of life. But landing on Mars is a risky business – NASA engineers experienced "seven minutes of terror" as they waited for the Curiosity rover to touch down safely in 2012.

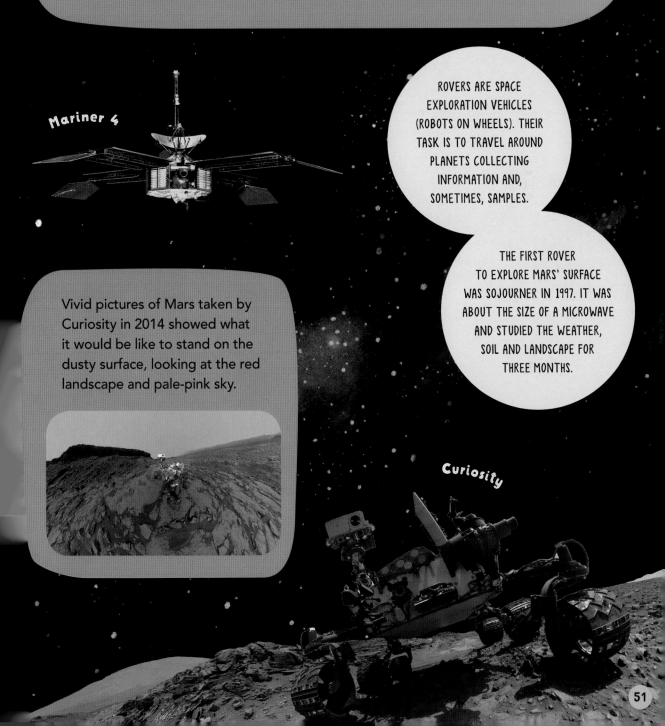

Mariner 4

ROVERS ARE SPACE EXPLORATION VEHICLES (ROBOTS ON WHEELS). THEIR TASK IS TO TRAVEL AROUND PLANETS COLLECTING INFORMATION AND, SOMETIMES, SAMPLES.

THE FIRST ROVER TO EXPLORE MARS' SURFACE WAS SOJOURNER IN 1997. IT WAS ABOUT THE SIZE OF A MICROWAVE AND STUDIED THE WEATHER, SOIL AND LANDSCAPE FOR THREE MONTHS.

Vivid pictures of Mars taken by Curiosity in 2014 showed what it would be like to stand on the dusty surface, looking at the red landscape and pale-pink sky.

Curiosity

MARS

ACCORDING TO NASA, THE FIRST PERSON WHO IS GOING TO LAND ON MARS IS ALIVE TODAY. IT COULD BE ME OR, MORE LIKELY, YOU!

A LANDER CALLED INSIGHT IS CURRENTLY ON MARS, EXAMINING THE PLANET'S INNER STRUCTURE. IT WILL ALSO LISTEN FOR "MARSQUAKES."

THE RED PLANET

We are now approaching my favorite planet – Mars. Of all the places in the Solar System that we are going to see, this is the visit I'm most excited about. Mars is the pinky, orange-red planet that you can see from Earth. The descent to the surface should be smooth compared to Venus, but we must still be careful – around 50% of visiting spacecraft have crashed while trying to land.

Why do I want to go to Mars so much? Well, for one thing, Mars is the planet that's most like Earth. As we get closer to the surface, look out for features that remind you of Earth – polar ice caps, mountains and dried-up riverbeds. They are easy to spot because the atmosphere is so thin.

During the visit, you can test your stamina by climbing Olympus Mons, the largest volcano in the Solar System. We'll also consider whether humans could one day live here. Now, put on your spacesuit and take an oxygen supply with you. The atmosphere is mainly carbon dioxide, which humans can't breathe.

MARS STATS

SIZE (DIAMETER)
6,792 km

AVERAGE DISTANCE FROM EARTH
78 million km

HOW MANY BIRTHDAYS?
A year lasts for 687 Earth days, so you'd only get a birthday once every two years.

HOW LONG IS A DAY?
24 Earth hours and 40 Earth minutes

TAKE IN THE VIEW

Mars has a surface we can stand on, but we still have to pick our landing site carefully. The first things you'll notice when you step outside are the desolate landscape, pink sky and red surface dust. You might also notice the wind, which can whip up the dust, coating everything in sight. Storms on Mars can get so severe that they fill the sky with dust and block out the Sun's light.

MARS' RED COLOR COMES FROM THE IRON OXIDE (RUST) IN ITS SURFACE.

Phobos

PHOBOS LOOKS A BIT LIKE A POTATO AND DEIMOS LOOKS LIKE A PEBBLE. EACH IS ABOUT THE SIZE OF A SMALL TOWN.

MARTIAN SKY

Today the winds are calm and the sky is clear. The Martian sky is similar to Earth's sky, except that it is light pink instead of blue. Both stargazing and moon watching may be possible tonight, so we should have a view of Phobos and Deimos, Mars' tiny moons.

Deimos

DEEP FREEZE

It can be much colder here than on Earth because Mars is farther from the Sun. Temperatures range from a chilly -143°C (almost ten times colder than your freezer) to a comfortable 20°C.

MARS HAS THE LARGEST KNOWN CANYON IN THE SOLAR SYSTEM. THE VALLES MARINERIS STRETCHES A WHOPPING 4,000 KM – LONG ENOUGH TO CROSS FROM THE US' EAST COAST TO THE WEST.

SEASONS ON MARS

Northern summer

Southern summer

Like Earth, Mars is tilted on its axis, so it also has seasons. They happen because different areas of the planet face toward or away from the Sun during its orbit. However, because Mars' tilt is slightly more extreme than Earth's (25° rather than 23.5°), and it has a longer year and a different orbit, the lengths of Martian seasons are not the same as ours. For instance, spring in Mars' northern hemisphere lasts for seven months. Dust storms can be particularly bad at that time of year, so be prepared.

OLYMPUS MONS

No trip to Mars would be complete without a visit to Olympus Mons, the largest volcano in the Solar System. Olympus Mons is an awe-inspiring 25 km tall – nearly three times the height of Mount Everest – and, at 624 km wide, it would cover most of France. Climbing Olympus Mons is not for the fainthearted, so bring climbing equipment and supplies.

OLYMPUS MONS IS A SHIELD VOLCANO. THE LAVA FROM A SHIELD VOLCANO FLOWS FOR LONG DISTANCES BEFORE SOLIDIFYING, FORMING A WIDE VOLCANO WITH GENTLE, SLOPING SIDES (GREAT FOR CLIMBING!). THE EDGE OF THE VOLCANO IS SURROUNDED BY A DRAMATIC 10-KM-HIGH CLIFF.

HOW BIG IS IT?

Have a look at how big Olympus Mons is compared to Earth's tallest mountains.

OLYMPUS MONS — 25,000 M

| 3,776 M MOUNT FUJI | 4,810 M MONT BLANC | 5,895 M MOUNT KILIMANJARO | 6,961 M ACONCAGUA | 8,848 M MOUNT EVEREST |

MARS' VOLCANOES ARE MUCH BIGGER THAN THE ONES ON EARTH. THIS IS PARTLY BECAUSE MARS' CRUST DOESN'T MOVE ABOUT IN THE WAY THAT EARTH'S DOES. THIS MEANS THAT THE LIQUID ROCK UNDERNEATH CAN ONLY ESCAPE FROM A FEW PLACES, CREATING VERY LARGE VOLCANOES.

CLIMBING OLYMPUS MONS

TIME
It would take several weeks to climb to the very top.

TERRAIN
A sheer cliff, then a steady climb to the summit.

EQUIPMENT
You'd need a flexible spacesuit, climbing ropes and oxygen.

TEMPERATURE
Typically -60°C

LIFE ON MARS

IN 2018, NASA'S CURIOSITY ROVER FOUND EVIDENCE OF ANCIENT ORGANIC MOLECULES — SOME OF THE MATERIALS NEEDED TO SUPPORT LIFE.

Almost 50 probes have been sent to Mars, but we still don't know if any life form existed on the planet or survives today. Was there ever life on Mars and will people be able to set foot on the planet in the future?

THE NEXT STEP

The dried-up riverbeds on Mars prove that there used to be liquid water on the planet. Some evidence suggests that water may still exist beneath the surface. NASA's next mission there, the Mars 2020 rover, will hopefully tell us more. The rover will study the Martian environment and collect samples that we hope can be brought back to Earth. NASA plans to see if it's possible to live on Mars and search for evidence of ancient life there.

A ROVER NAMED EXOMARS IS ALSO BEING DEVELOPED BY THE EUROPEAN AND RUSSIAN SPACE AGENCIES. ONE OF ITS FUTURE TASKS WILL BE TO DIG BELOW THE MARTIAN SURFACE TO LOOK FOR SIGNS OF LIFE.

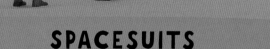

SPACESUITS

Any future humans on Mars would need to wear a spacesuit whenever they are outside to protect them from the high levels of CO_2 in the atmosphere, radiation from the Sun, huge changes in temperature and sudden dust storms.

HUMANS ON MARS

We know that Mars has similar features to Earth, but Mars is also very different in ways that would make it hard for life to survive. So, what sort of things do we need to consider before sending a human to Mars?

ONE OF THE MAIN THINGS STOPPING HUMANS FROM GOING TO MARS AT THE MOMENT IS THE HUGE COST THAT WOULD BE INVOLVED. MOST SPACE AGENCIES AGREE THAT A HUMAN WON'T SET FOOT ON MARS UNTIL THE LATE 2030s.

BUILDING

To live on Mars, future explorers will need to build protected, pressurized structures on the planet. This building (left) shows what a future house on Mars could look like. It would provide a breathable atmosphere in which to live and grow food. Scientists are currently conducting experiments to see which types of plants might be able to grow there. They are also thinking about designs for 3-D-printed houses.

RETURN JOURNEY

Human explorers will need a spacecraft that is able to take off from Mars' surface, so they can return to Earth. But some missions are already being planned with no return spacecraft. Would you be interested in a one-way trip to Mars?

NEXT STOP! ➡

It's time to continue our journey, but maybe you will come back here for real one day.

ASTEROID BELT

We have left Mars and are now heading toward a gigantic expanse of rocks that are in orbit around the Sun. This region, called the asteroid belt, lies on the way to Jupiter, the next planet on our tour. As we pass through, we need to take care – asteroids do sometimes collide with each other and change direction. Keep alert and let me know if you notice any rocks getting too close to our spaceship.

BUMPY RIDE

We might experience slight turbulence as we travel to the other side of the belt. This is caused by the gravitational pull of the largest bodies found here. Keep your seat belt fastened as we will need to swerve to avoid the smaller rocks, too.

ROCK COMPOSITION

Asteroids are mostly made of rock, metal and other matter. Many also contain water. They vary hugely in shape, weight and size – most are pebble sized or a few meters wide, but some can be hundreds of kilometers in diameter.

FAR FROM HOME

Most asteroids in our Solar System sit in the main belt between Mars and Jupiter, but some exist outside it. If any of these wanderers get close to Earth we call them "near-Earth objects." Scientists monitor them closely to make sure they don't crash into our home planet.

MINING FOR METALS

In the future, we may want to mine asteroids for metals and minerals that are in short supply on Earth. There are already plans to send missions to the asteroid belt to investigate this idea. This is what a mining site could look like.

Ceres

BIGGEST ROCK ON THE BLOCK

Keep an eye out for Ceres, the largest rock in the belt. Ceres is so large that in 2006 it was promoted from an asteroid to a dwarf planet. It is the closest dwarf planet to Earth that has been found so far – all others sit out in the Kuiper Belt (pages 90-91). Ceres was visited by spacecraft Dawn in 2015. It flew past the dwarf planet and sent back spectacular pictures of its surface.

ROCKY BODIES

From planets to comets and everything in between, there are many different types of bodies floating around the Solar System. Although they may look similar, each type has its own defining features. Let's take a closer look ...

Dirty Snowball

DIRTY SNOWBALLS

Keep your eyes peeled for icy rocks whizzing past. These are comets and are made of water, frozen gases, rock and dust. Comets travel around the Sun in large, elliptical orbits. When they pass near the Sun, the solar wind (pages 98-99) causes them to heat up and release dust and gases. This can form a tail that stretches away from the Sun.

SOME ASTEROIDS ARE LARGE ENOUGH TO HAVE THEIR OWN MOON. ASTEROID 87 SYLVIA IS THE EIGHTH-LARGEST ASTEROID IN THE BELT AND THE ONLY ONE CURRENTLY KNOWN TO HAVE TWO MOONS — ROMULUS AND REMUS.

AMAZING ASTEROIDS

Asteroids are thought to be the rubble that was left over after the main planets of the Solar System formed. This means that they can help us understand what the early Solar System was like. Each one is a time capsule allowing us to look back roughly 4.5 billion years.

METEOROID, METEOR OR METEORITE?

These names are all very similar, so what exactly is the difference between them? They are actually the same things but at different stages of their journeys down to the Earth's surface.

The Sun

Meteoroid

METEOROID

A meteoroid is a small chunk of rock, often a broken-off part of an asteroid, in orbit around the Sun.

METEOR

If a meteoroid gets knocked out of its orbit and hits Earth's atmosphere, it becomes a meteor. Friction between the atmosphere and the rock makes it superhot, causing it to glow. Meteors are also called shooting stars.

METEORITE

A meteorite is a meteor that is large enough to have survived its passage through Earth's atmosphere and landed on the planet's surface.

Meteor

Meteorite

EXPLORING THE
OUTER SOLAR SYSTEM

We are about to leave the asteroid belt and enter the outer Solar System, a region beyond our neighboring planets that's home to the gas and ice giants – Jupiter, Saturn, Uranus and Neptune. Before we do, let's take a look at some of the missions that have already traveled out here and some of their astounding discoveries.

1972

PIONEER 10

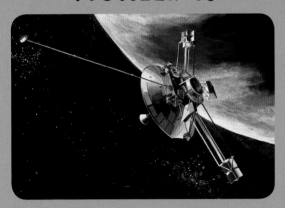

In 1972, Pioneer 10 was the first spacecraft to leave the inner Solar System and travel past the asteroid belt. It sent back the first pictures of Jupiter and measured its magnetic field, atmosphere and interior.

Pioneer 11 followed a year later and investigated Saturn.

1977

THE VOYAGERS

Voyager 1 and Voyager 2 were launched in 1977. The two-craft mission studied all four outer planets – Jupiter, Saturn, Uranus and Neptune.

The spacecraft also collected data from the farthest parts of the Solar System. They are currently exploring interstellar space (pages 102-103).

DEEP SPACE NETWORK

We send instructions to, and receive information from, space probes using radio waves. They can travel huge distances through space but can become hard to detect when very far away.

That's where NASA's Deep Space Network comes in. The network is made up of three groups of giant radio antennae (devices used to send and receive radio signals). The antennae need to be big to receive the weak signals from a distant space probe. They are located in Spain, Australia and the US, so that, no matter where a probe is in the Solar System, there is an antenna on Earth that can receive its signal.

1997

CASSINI-HUYGENS

The Cassini-Huygens mission, formed of the Cassini spacecraft and Huygens probe, was launched in 1997 and orbited Saturn for 13 years. Its task was to study Saturn's ring system and moons. The Huygens lander also touched down on Saturn's largest moon, Titan. Find out more on pages 80-81.

2004

ROSETTA

The Rosetta probe was launched in 2004 and flew out beyond the asteroid belt. Its mission was to study a comet named 67P/Churyumov-Gerasimenko. The main probe orbited the comet and then released Philae, a small lander. Philae bounced across the comet's surface ... and landed nearly upside down against a cliff. Luckily, there was little damage, but its solar panels, which needed sunlight to work, were in shadow. Philae was able to transmit valuable information about the comet's structure back to Earth before running out of power.

FLYING ABOARD VOYAGERS 1 AND 2 ARE TWO IDENTICAL RECORDS CARRYING GREETINGS IN 60 LANGUAGES, SAMPLES OF MUSIC FROM DIFFERENT CULTURES AND ERAS, AND NATURAL AND HUMAN-MADE SOUNDS FROM EARTH.

JUPITER

EARTH'S PROTECTOR

Our first stop in the outer Solar System is a stunning sight. Jupiter is two and a half times bigger than any other planet and by far the largest in our Solar System. Over 1,300 planet Earths could be squished inside it! I call Jupiter "Earth's bodyguard" because it protects Earth from some of the space rocks that might hit it. Jupiter's humungous mass gives it such a powerful gravitational pull that it draws these space rocks toward it and away from Earth.

Even though it is very dense, Jupiter doesn't have a solid surface that we can land on. Instead, it has a thick, cloudy atmosphere. Exploring these colorful, swirling bands will be an awesome experience. Hold on tight – Jupiter is a windy place so it's likely to be a bumpy ride.

JUPITER IS THE PLANET THAT IS MOST SIMILAR TO THE SUN – BOTH ARE MADE OF HYDROGEN AND HELIUM.

MANY SPACECRAFT HAVE TRAVELED TO JUPITER AND OTHER SPACECRAFT HAVE USED ITS STRONG GRAVITY AS A CATAPULT TO SEND THEM OUT INTO DEEP SPACE.

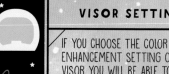

VISOR SETTING

IF YOU CHOOSE THE COLOR ENHANCEMENT SETTING ON YOUR VISOR YOU WILL BE ABLE TO SEE THE SWIRLS OF JUPITER'S STORMS IN AMAZING DETAIL.

JUPITER SPINS AROUND
FASTER THAN ANY OTHER
PLANET IN THE SOLAR SYSTEM —
IT ROTATES ONCE EVERY TEN HOURS.
SPINNING SO QUICKLY CAUSES
THE PLANET TO BULGE
OUT IN THE MIDDLE.

JUPITER STATS

SIZE (DIAMETER)
142,984 km

AVERAGE DISTANCE
FROM EARTH
628 million km

HOW MANY BIRTHDAYS?
One every 12 Earth years

HOW LONG IS A DAY?
Ten Earth hours

THROUGH THE CLOUDS

As we approach Jupiter, the faint rings that surround it come into view – it's as if they've suddenly appeared out of nowhere! The cloudy surface swirls mysteriously and the Great Red Spot peers up at us from below. Let's dive down for a close-up look at the planet's turbulent cloud systems and its unusual structure.

IF A PERSON COULD STAND ON JUPITER'S CLOUDS, THE FORCE OF GRAVITY WOULD BE ABOUT 2.4 TIMES THAT ON THE SURFACE OF EARTH. SO A PERSON WHO WEIGHS 45 KG ON EARTH WOULD WEIGH ABOUT 90 KG ON JUPITER! THIS WOULD MAKE MOVING VERY DIFFICULT!

JUPITER'S LAYERS

CLOUD

Falling through the thick, cloudy atmosphere of swirling orange and white, the temperature begins to drop and the strong winds pick up.

LIKE THE PLANETS OF THE INNER SOLAR SYSTEM, YOU CAN SEE JUPITER FROM EARTH WITHOUT A TELESCOPE ON A CLEAR NIGHT.

GAS

The zones of cloud are made of different mixes of gases, which is why they are different colors. Belts of cloud that are next to each other have winds that move them in different directions.

OCEAN

Farther down, there's a surprise – the largest ocean in the Solar System! It is made of fluid hydrogen, something that's not quite gas but also not quite liquid – a very peculiar substance.

JUPITER'S CLOUDS ARE THOUGHT TO BE ABOUT 50 KM THICK. WE THINK THE DARKER CLOUDS ARE FORMED FROM CHEMICALS BROUGHT UP FROM BELOW WHICH CHANGE COLOR WHEN THEY REACT WITH SUNLIGHT.

METAL CORE

At Jupiter's center the pressure becomes so great that the hydrogen is squeezed into a part-liquid, part-solid, metal-like substance. It is this core that gives the planet its powerful magnetic field, which is 20,000 times stronger than Earth's.

THE GREAT RED SPOT

Get your camera ready – we're heading toward one of the Solar System's most famous sights. A giant storm, named the Great Red Spot, has been raging in Jupiter's atmosphere for more than 300 years. It is so colossal that it measures TWICE THE WIDTH of Earth. Scientists think that the spot's color is created by the gases inside reacting with ultraviolet rays from the Sun. But much about the storm remains a mystery. This stupendous, spinning vortex is like a hurricane, so we'll observe from a safe distance.

JUST LIKE STORMS ON EARTH, THE GREAT RED SPOT HAS A STILL CENTER, CALLED THE "EYE."

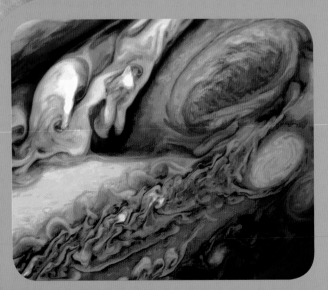

SISTER STORMS

Many other storms rage across Jupiter. Some are white and have become known as the "string of pearls." These storms come and go. Since 1986, the number of visible "pearls" has varied from six to nine.

WHIRLING WINDS

Winds in the Great Red Spot can reach speeds of up to 650 km per hour. The storm spins counterclockwise and takes six Earth days to spin around once.

THE SHAPE OF THE GREAT RED SPOT IS CHANGING. IT SEEMS TO BE SHRINKING AND IT IS NOW HALF THE SIZE THAT IT WAS 100 YEARS AGO.

JUNO MISSION

In 2011, a mission called Juno was launched to investigate Jupiter. Its aim is to look below the swirling outer clouds of the planet to discover exactly what lies beneath. Scientists hope to get a better understanding of how this mighty planet was formed, which may give them clues about how other planets came into existence, too. Juno arrived at Jupiter in 2016 and has been sending back some incredible images of the planet ever since.

JUPITER'S MOONS

Jupiter has a LOT of moons – 79 have been discovered so far, but more may still be detected. In 2018 alone, ten new moons were found. Let's visit the four largest. These are known as the Galilean moons, named after the astronomer Galileo Galilei, who first saw them through his telescope.

EUROPA has a crust of ice that scientists think could be hiding a salty ocean beneath. The ice is thought to be 16-24 km thick. Europa is one of the few places in the Solar System where life could exist due to the presence of this liquid water.

EUROPA IS THE SMALLEST OF THE GALILEAN MOONS.

3,121 km wide – that's a bit smaller than Earth's Moon.

IO has over 400 live volcanoes, making it one of the most actively volcanic places in the Solar System. Other moons in the outer Solar System have an icy crust, but Io's rocky surface has a thin coating of frozen sulfur dioxide and lakes of molten silicate lava.

IO IS CLOSEST TO JUPITER.

3,660 km wide

GANYMEDE is so large that it is even bigger than Mercury. The dark regions on its surface are covered in craters, which tells us that the moon's surface is very old. The light areas are thought to be where water has leaked onto the surface and frozen, leaving a smooth finish.

GANYMEDE IS JUPITER'S LARGEST MOON.

5,268 km wide

CALLISTO has a heavily cratered surface. Scientists think a saltwater ocean could lie beneath the icy crust. Oxygen has been detected in the exosphere (the outermost layer of the atmosphere).

EVIDENCE OF WATER ON CALLISTO AND EUROPA MEANS THAT LIFE MAY EXIST ON THESE MOONS.

4,821 km wide

SATURN

THE LORD OF THE RINGS

We're now a VERY long way from the Sun, and pale-yellow Saturn is a breathtaking sight in the darkness. Because the planet's icy rings are so thin, they appear and disappear from view as our craft changes direction, so watch carefully.

Like Jupiter, Saturn is a gas giant, which means it doesn't have a solid surface. Instead of landing, let's take a tour of the amazing ring system, followed by a flyby of Saturn's many moons. There will be plenty of photo opportunities, so have your camera at the ready. The geysers erupting from the surface of Enceladus, Saturn's sixth-largest moon, are an especially awesome sight.

Our spacecraft is one of the few to travel this far into the Solar System. Before we leave, you can learn more about another important mission, the Cassini-Huygens space probe, which reached Saturn back in 2004.

SATURN STATS

SIZE (DIAMETER)
120,536 km

AVERAGE DISTANCE FROM EARTH
1.2 billion km

HOW MANY BIRTHDAYS?
You would have to wait 29 Earth years to celebrate your first birthday on Saturn!

HOW LONG IS A DAY?
10.7 Earth hours

THE ANGLE AT WHICH WE VIEW THE RINGS FROM EARTH CHANGES AS SATURN ORBITS. SOMETIMES THEY SEEM TO DISAPPEAR COMPLETELY!

SATURN IS THE LEAST-DENSE PLANET IN THE SOLAR SYSTEM. IT IS LESS DENSE THAN WATER. IF YOU HAD A BATHTUB BIG ENOUGH, SATURN WOULD FLOAT IN IT!

SATURN'S RINGS

Hold on to your helmet and prepare to fly through the biggest and brightest rings in the Solar System. Jupiter, Uranus and Neptune also have rings, but Saturn's system is by far the most spectacular. Although only about 10 meters deep, it's a whopping 280,000 km wide.

SATURN HAS SEVEN MAIN RING GROUPS. IT WOULD TAKE AROUND TEN DAYS TO DRIVE THROUGH SOME OF THEM IN A CAR.

THERE ARE GAPS BETWEEN THE RINGS. ONE GAP, CALLED THE CASSINI DIVISION, IS 4,800 KM WIDE.

SATURN'S SURFACE?

We can't land on the surface – it's just a mass of gases – but there's plenty to see once we've passed through the rings. Like Jupiter, Saturn has ferocious storms. One of the most stunning sits at Saturn's north pole. The eye of this gigantic hurricane is estimated to measure 2,000 km across and has cloud speeds of 530 km per hour (the fastest hurricanes on Earth only reach speeds of about 250 km per hour).

EACH RING ORBITS SATURN AT A DIFFERENT SPEED, DEPENDING ON ITS DISTANCE FROM THE PLANET.

INSIDE THE RINGS

As we get closer, we can see that the rings are made of billions of different-sized objects, from tiny specks of dust to giant, icy boulders as big as mountains. Each individual object is in orbit around Saturn. The objects are mainly made of water ice, which reflects the weak light of the Sun beautifully, and a little bit of rock. Scientists think the objects in the rings are the remains of comets, asteroids and moons, broken up by the planet's fierce gravity.

MOONS
AND MOONLETS

Scientists have discovered 62 moons orbiting Saturn so far – see if you can spot any new ones as we fly past. Saturn's moons come in all sorts of shapes and sizes. They range from giant Titan, which is larger than Mercury, to minute Mimas, which is about eight times smaller than Earth's Moon. Here are some highlights to look out for.

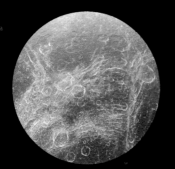

DIONE

Saturn's fourth-largest moon is covered in sparkling ice cliffs.

PHOEBE

A dark, mysterious moon, scientists think Phoebe may be a captured "centaur"– an ancient body from the Kuiper Belt that's traveled into the Solar System.

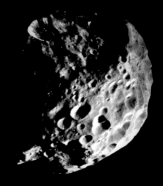

IAPETUS

This moon has a distinctive two-toned terrain – half the surface is black and the other half is white.

PAN AND ATLAS

Pan and its neighbor, Atlas, are tiny moons that look like pieces of filled pasta. Their unusual shape is thought to be due to collisions with other objects.

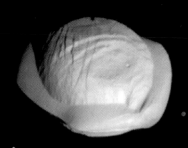

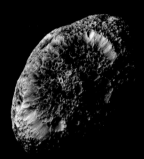

HYPERION

This is the largest of Saturn's irregular moons. It looks a bit like a giant sea sponge.

BABY MOONS

Shining clumps of ice sit in the gaps between Saturn's seven rings. These icy clusters are "moonlets" (baby moons). Scientists believe the moonlets keep the rings in place.

THE MOONLETS RANGE FROM THE SIZE OF A TRUCK TO THE SIZE OF A SPORTS STADIUM.

CASSINI:
MISSION TO SATURN

One of the few spacecraft to have made it as far as Saturn was the Cassini-Huygens space probe, which launched in October 1997. It was formed of two parts – the Cassini spacecraft that orbited Saturn and the Huygens probe that explored Titan (Saturn's largest moon).

The spacecraft took seven years to fly from Earth to Saturn. It survived an astounding 13 further years in Saturn's orbit and discovered six new moons during that time. The mission ended in 2017, when the craft was deliberately crashed into Saturn's atmosphere. The spacecraft was low on fuel and a controlled crash was the only way it would meet a safe end.

HUYGENS WAS THE FIRST PROBE TO TOUCH DOWN ON A BODY IN THE OUTER SOLAR SYSTEM.

SATURN'S SEASONS

Cassini sent back images of a six-sided jet stream (a system of air currents) at Saturn's north pole. Over four years, the slowly changing colors of this vast weather pattern moved from shades of blue to gold. The varying colors are believed to be linked to Saturn's seasons.

EXPLORING TITAN

The Huygens lander was released from Cassini on December 25, 2004. It took time-lapse images as it fell through Titan's atmosphere and reached the surface two and a half hours later.

We were able to look through Titan's thick, cloud-filled atmosphere for the first time. The lander made several exciting discoveries. Earthlike features were found on Titan, such as rain, rivers, lakes and seas. But instead of liquid water, they were made up of liquid methane (the gas that cows on Earth – and humans – produce when they pass gas!).

ENCELADUS IS ONLY 504 KM ACROSS. THAT'S SMALLER THAN THE STATE OF ARIZONA.

EXPLORING ENCELADUS

Data sent back from Cassini also revealed astonishing facts about Enceladus, one of Saturn's other moons.

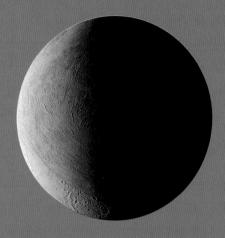

Jets of liquid water were shooting hundreds of kilometers out into space through cracks in Enceladus' surface.

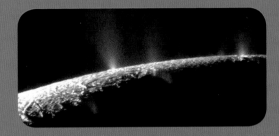

A huge, liquid-water ocean was discovered below its icy crust. The plumes of water erupting from the surface contained complex carbon molecules. This was a REALLY exciting discovery. It is evidence that hydrothermal (hot-water) vents might exist on this moon's ocean floor. Like the hydrothermal vents in Earth's oceans, the conditions to host life could also exist in this distant corner of the outer Solar System.

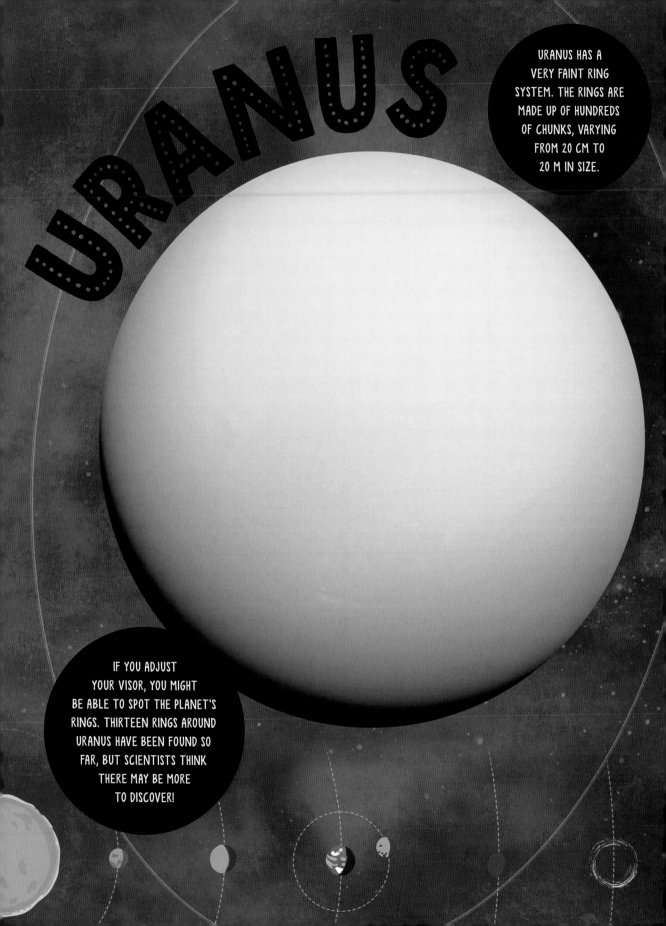

URANUS

URANUS HAS A VERY FAINT RING SYSTEM. THE RINGS ARE MADE UP OF HUNDREDS OF CHUNKS, VARYING FROM 20 CM TO 20 M IN SIZE.

IF YOU ADJUST YOUR VISOR, YOU MIGHT BE ABLE TO SPOT THE PLANET'S RINGS. THIRTEEN RINGS AROUND URANUS HAVE BEEN FOUND SO FAR, BUT SCIENTISTS THINK THERE MAY BE MORE TO DISCOVER!

A PLANET
ON ITS SIDE

Ahead lies Uranus, a beautiful, aquamarine sphere that hangs serene and still in the cold darkness. Unlike any other planet in the Solar System, Uranus spins on its side. It's possible that a huge collision knocked it over soon after it was formed.

Uranus is an "ice giant" – a huge planet similar to the gas giants, only made from heavier stuff. Other than us, just one spacecraft, Voyager 2, has flown past Uranus. So we have the chance to do something no human has done before – see the planet up close. This visit is going to be EXCITING!

We're going to kick off with a descent through the ice-cold atmosphere, followed by a plunge into a strange slush ocean. Then, we'll head to Miranda, one of Uranus' moons, for some extreme sports – low-gravity parachuting from the tallest cliff in the Solar System.

URANUS HAS 27 MOONS. THE PLANET HAS A STRONG GRAVITATIONAL PULL, WHICH MAY HAVE CAPTURED SOME OF THESE MOONS FROM THE NEARBY KUIPER BELT.

URANUS STATS

 SIZE (DIAMETER)
51,118 km

 AVERAGE DISTANCE FROM EARTH
2.7 billion km

 HOW MANY BIRTHDAYS?
A year is 84 Earth years; that's a long time to wait for your first birthday!

 HOW LONG IS A DAY?
17 Earth hours

SLUSHY SURFACE

Our craft is passing through the outer layers of Uranus' atmosphere. A pale, blue gas – mostly hydrogen and helium – surrounds us. We're in the coldest atmosphere in the entire Solar System. Temperatures here can drop as low as -224°C. That's more than FIVE TIMES colder than the North Pole in winter.

As you descend, the atmosphere becomes thick and cloudy with water, ammonia and methane, and it starts to get warmer.

We fall farther down toward the planet's core and the atmosphere gets so thick that it turns into a slush ocean! The charged particles whizzing around in this ocean are thought to influence Uranus' magnetic field.

THE REASON URANUS' ATMOSPHERE IS SO COLD IS THOUGHT TO BE BECAUSE THE PLANET HAS A SURPRISINGLY LOW CORE TEMPERATURE COMPARED TO OTHER PLANETS. THIS MIGHT BE CAUSED BY URANUS SPINNING ON ITS SIDE.

The pressures are so high in the depths of the slush that our craft may be pelted by "diamond rain." This is because the extreme pressure squeezes the carbon in the slush ocean, turning it into diamonds that fall toward the planet's core. OUCH!

LIKE ON JUPITER,
STORMS SOMETIMES APPEAR
IN URANUS' ATMOSPHERE. THEY
CAN REACH WIND SPEEDS OF UP
TO 900 KM PER HOUR. BETTER
HOLD ON TO YOUR HELMET!

The atmospheric pressure means that we can't stay here too long. Let's head to one of Uranus' moons, which has a surface we can land on.

VERONA RUPES

We've left Uranus' swirling atmosphere behind and landed on Miranda, Uranus' fifth-biggest moon. Not only that, we've touched down at the top of the tallest cliff in the Solar System – Verona Rupes. This vast, icy cliff rises 20 km above Miranda's surface – that's ten times the depth of the Grand Canyon on Earth.

Miranda has ten times less mass than our Moon, so the force of gravity here is very low. Low gravity = FUN! Because gravity is so weak here, if you parachuted off the top of Verona Rupes, it would take around ten minutes to reach the bottom.

NEPTUNE

A TURBULENT PLANET

We are now 30 AU from the Sun (30 times the distance from Earth to the Sun) and approaching cool, blue Neptune. If you look closely, you might be able to spot its very faint ring system.

Away from the hustle and bustle of the inner Solar System, you'd expect Neptune to be a peaceful place. However, violent winds – some of the strongest in the Solar System – are blowing in its atmosphere and things could get bumpy if we go too close. We can't land as there is no solid surface, but a flyby of Triton, Neptune's largest moon, has been arranged for us.

It takes five-and-a-bit days for Triton to revolve around Neptune, so there won't be time to witness its unusual backward orbit. However, we can make the most of our visit by observing its volcanic geysers.

NEPTUNE IS CURRENTLY CONSIDERED TO BE THE FARTHEST PLANET FROM THE SUN. BODIES THAT LIE BEYOND NEPTUNE'S ORBIT ARE KNOWN AS TRANS-NEPTUNIAN OBJECTS.

THE TILT OF
NEPTUNE'S AXIS IS
SIMILAR TO EARTH'S,
SO THE PLANET ALSO
EXPERIENCES
SEASONS.

NEPTUNE STATS

SIZE (DIAMETER)
49,528 km

AVERAGE DISTANCE
FROM EARTH
4.3 billion km

HOW MANY BIRTHDAYS?
One every 165 Earth years!

HOW LONG IS A DAY?
16 Earth hours

NEPTUNE IS
THE ONLY PLANET
THAT'S NOT VISIBLE
TO THE NAKED EYE
FROM EARTH.

THE LAST PLANET?

Our craft is now a long, long way out in the Solar System. Only Voyager 2 has traveled this far, when it flew past Neptune and Triton in 1989. Let's follow the same route for a close-up view of this mysterious ice giant and its most fascinating moon.

DEEP BLUE

Neptune's atmosphere is made up of a mixture of hydrogen, helium and methane gas. There isn't a solid surface on the planet. Instead, its incredibly deep atmosphere extends down and down, right down to the planet's rocky core. Scientists think the heat and pressure inside Neptune is high enough for its core to be surrounded by a liquid diamond ocean.

A GIANT VORTEX AS LARGE AS EARTH, CALLED "THE GREAT DARK SPOT," WAS VIEWED BY VOYAGER 2 IN 1989. IT HAS NOW DISAPPEARED FROM THE SURFACE, BUT KEEP A LOOKOUT FOR NEW STORMS AS WE FLY PAST.

COLD AND HOT

Neptune is unusual. Colder than Uranus, it is the ONLY planet in our Solar System to give out more heat than it receives from the Sun. This means that there must be a heat source coming from the planet's interior, and this may cause the windstorms raging in its atmosphere.

WE CURRENTLY THINK THAT NEPTUNE HAS 14 MOONS (13 CONFIRMED AND 1 AWAITING CONFIRMATION).

OUT HERE, THE SUN IS SO FAR AWAY THAT IT LOOKS LIKE A SUPERBRIGHT STAR. IF WE VISITED NEPTUNE AT LUNCHTIME — THE BRIGHTEST PART OF THE DAY ON EARTH — IT WOULD BE AS DARK AS LATE EVENING.

TRITON FLYBY

Triton's frozen surface has an icy sheen, which twinkles in our ship's headlights. It has rocky features but very few impact craters. The fact that the surface looks so smooth suggests that the moon is probably only a few million years old.

It's also thought to be highly active. There goes a geyser! Eruptions like this can send long plumes of nitrogen spewing as high as 8 km into the air.

THE KUIPER BELT:
A FROZEN DISK

Ahead of us lies the Kuiper Belt – a VAST expanse of hundreds of thousands of icy bodies. It sits like a doughnut around the other objects in the Solar System and stretches 20 to 30 astronomical units across space. Pluto and three more dwarf planets – Eris, Haumea and Makemake – are found here. We're going to make two stops, on Haumea and Pluto, before leaving this freezing region. Be warned: fast-spinning Haumea might make you feel a little dizzy …

NEW HORIZONS

The New Horizons space probe was launched in 2006 to study Pluto and other objects in the Kuiper Belt. It reached an object deep in the belt, called "2014 MU69" or Ultima Thule, on January 1, 2019. This is the first small Kuiper Belt object ever to be explored by a spacecraft.

LIKE THE EIGHT PLANETS IN THE SOLAR SYSTEM, DWARF PLANETS ORBIT THE SUN AND ARE NEARLY ROUND IN SHAPE. HOWEVER, DWARF PLANETS ARE MUCH SMALLER THAN PLANETS. BECAUSE DWARF PLANETS ARE SMALL, THEY ARE NOT ABLE TO CLEAR THEIR ORBITS OF OTHER OBJECTS, AS TRUE PLANETS DO.

LIGHT CURVES

Most Kuiper Belt objects are too small and far away for telescopes to see in detail. So scientists have learned about them by creating light curves – graphs that plot the light an object reflects out.

Imagine we spot a far-off, unknown object. The only thing we're certain about is that the object rotates. We plot the light it reflects over several days and notice that it doesn't change. This tells us that the object is partly spherical in shape because a spherical object will reflect the same amount of light as it rotates. A sausage-shaped object would reflect varying levels of light as it rotates, depending on whether its short or long sides were facing us, so its light curve would look different.

IT'S THOUGHT THE KUIPER BELT IS MADE UP OF PIECES LEFT OVER AFTER THE FORMATION OF THE SOLAR SYSTEM. STUDYING IT CAN GIVE US AN UNDERSTANDING OF WHAT THE EARLY SOLAR SYSTEM WAS LIKE.

HAUMEA

Haumea is a fast-spinning, egg-shaped dwarf planet. Its high-speed rotation explains the squashed shape – the force of the spin pushes Haumea's matter outward, so it is fatter around the middle.

PLUTO WAS DISCOVERED BY CLYDE TOMBAUGH, AN ASTRONOMY ASSISTANT, IN 1930. IT WAS INITIALLY THOUGHT TO BE LARGER THAN MERCURY.

PLUTO IS THE ONLY ASTRONOMICAL BODY NAMED BY AN 11-YEAR-OLD. A BRITISH GIRL, VENETIA BURNEY, CAME UP WITH THE NAME WHEN PLUTO WAS FIRST DISCOVERED. OTHER SUGGESTIONS WERE ZEUS, CRONUS AND PERCIVAL.

PLUTO STATS

SIZE (DIAMETER)
2,370 km

AVERAGE DISTANCE FROM EARTH
5.7 billion km

HOW MANY BIRTHDAYS?
A year is 248 Earth years – birthdays don't exist here!

HOW LONG IS A DAY?
6.9 Earth days

PLUTO

A DWARF PLANET

Do you like winter sports and getting away from it all? If your answer is yes, then Pluto is the place for you!

No trip to the Kuiper Belt would be complete without a visit to its best-known dwarf planet. In the early 1900s, scientists thought there could be another planet beyond Neptune and Pluto was eventually found in 1930. It was originally named the Solar System's ninth planet, but years later, after improvements in space technology led to the discovery of similar-sized Eris, Haumea and Makemake, it was reclassified as a dwarf planet.

Take a look at the frozen surface as we approach. The landscape is rugged and mountainous, and the icy cliffs sparkle in our ship's headlights. Unlike the last few planets, we can land here – let's look for a good place to land so we can stretch our legs.

PLUTO IS TINY. EIGHT PLUTOS COULD FIT COMFORTABLY INSIDE MERCURY, THE SOLAR SYSTEM'S SMALLEST PLANET.

39.5 AU FROM THE SUN

A FROZEN WORLD

As we get closer, see if you can spot a pale, heart-shaped area – the Tombaugh Region – covering a large part of the frozen surface. Most of Pluto is spotted with craters, but the left half of this region – named Sputnik Planitia – is smooth and will make a perfect place to set down our spacecraft.

 FROSTBITE WARNING!

The cold here is EXTREME. It's around -230°C at the surface, so it's vital we regulate our spacesuits' temperatures before disembarking. Don't under ANY circumstances remove your space gloves. The light level is lower than we're used to, so we'll take a flashlight as well.

NOT MUCH WAS KNOWN ABOUT PLUTO BEFORE THE 2015 FLYBY MISSION OF THE NEW HORIZONS PROBE. INSTEAD OF A DULL, ICY ROCK, IT REVEALED A FASCINATING, FROZEN WORLD.

RED SNOW?

As well as being icy, Pluto's surface is surprisingly colorful – notice the variety of white, yellow and orange features as we come in to land. You're likely to spot strange dark-red patches covering some areas, too. This gunky substance is called tholin and it is formed from a reaction between methane gas in Pluto's atmosphere and the Sun's ultraviolet light.

FASCINATING FEATURES

Pluto's terrain is a snowboarder's dream. Its natural features include craters, valleys and amazing frozen mountains. These giant blocks of water ice rise as high as 3,500 meters. Pluto's gravity is low – around 8% of Earth's – which means you can do some epic jumps!

MOONS

As we leave Pluto, we'll carry out a flyby past the planet's five moons – Charon, Styx, Nix, Hydra and Kerberos. Charon, the largest, is a similar size to Pluto. Can you see the red patches at its north pole? Scientists think that these are formed when gases in Pluto's atmosphere travel to Charon and get trapped at its pole, where they turn into tholins.

Charon

Styx   Nix   Hydra   Kerberos

PLANET NINE

Far beyond the Kuiper Belt, at the dark, icy edges of our Solar System, astronomers believe there may be another, undiscovered planet. Although nobody has actually seen it yet, scientists believe it exists because of the effect it has on the orbits of objects in the Kuiper Belt. The first person to see this mysterious planet may get the honor of naming it. What would you call it?

A SUPER-EARTH?

The evidence found by scientists so far suggests that this new planet – known for now as Planet Nine – is a super-Earth. This means it is bigger than Earth but smaller than the gas giants. It could be rocky, like Earth or Mars, or an icy gas planet like Neptune.

FAR FROM THE SUN

The planet's elliptical orbit is very stretched out. The closest it would come to the Sun is around 200 AU and the farthest is 1,200 AU.

A LONG ORBITAL PERIOD

The time it takes to orbit the Sun is estimated at 10,000 to 20,000 Earth years – that's really bad for the birthday count!

THE SEARCH FOR PLANET NINE

If it exists, finding Planet Nine will be incredibly exciting. It may challenge how we think the Solar System formed and has evolved.

LOCATION

Scientists think Planet Nine may have formed closer to the Sun before the gravitational pull of the larger planets threw it outward to its current location.

SEEING IT

Telescopes across the world have joined the search for Planet Nine, but a direct viewing will depend on where it is in its very long orbit. Viewing the planet is also tricky because it is so far away and faint – if it's in a part of the sky with lots of stars, it will be hard to pick out.

TRACKING OTHER OBJECTS

Until there's a sighting, scientists will continue to track the movement of smaller objects in the Kuiper Belt. This may help to pin down the planet's position.

WHERE DOES THE
SOLAR SYSTEM END?

The end of the Solar System isn't marked with a clear boundary. Because the Sun dominates our Solar System, some scientists think that its edge should be defined as the place least affected by the Sun's influences – namely, its gravitational and magnetic fields and the solar wind.

THE SOLAR WIND

Burning temperatures in the Sun's upper atmosphere generate superhot streams of high energy particles. These flow out into space at speeds of up to 400 km per second and create a vast bubble that extends around the Solar System. This bubble is called the heliosphere.

HELIOPAUSE

At the edge of the heliosphere, where we are now, the solar wind's power has become weak and the Sun's gravitational and magnetic fields no longer have much influence.

The weaker solar wind particles along the edge of the heliosphere are now affected by different, stronger forces. These forces are stellar winds and they come from stars far out in space.

THE MAGNETIC FIELD

The power of the Sun's magnetic field causes solar flares and plasma to erupt from its surface, sending matter and energy into the Solar System.

THE GRAVITATIONAL FIELD

The Sun's gravitational pull spreads right across the Solar System, keeping the eight planets and other bodies in orbit around it. The farther from the Sun we travel, the weaker the gravitational field gets.

THE EIGHT PLANETS (AND PLUTO)

KUIPER BELT

SOLAR WIND

Scientists call this frontier region the heliopause. It makes a good marker for the end of our Solar System.

ALTHOUGH THE SOLAR WIND IS INVISIBLE, WE CAN SEE THE EFFECTS OF IT FROM EARTH. WHEN THE CHARGED PARTICLES IN THE SOLAR WIND HIT THE EARTH'S ATMOSPHERE, THEY CAUSE THE SPECTACULAR NORTHERN AND SOUTHERN LIGHTS.

THE OORT CLOUD

We're nearing the end of our epic tour and our final destination is in sight. Ahead of us stretches a gigantic zone of icy objects, believed to surround the entire Solar System like an enormous sphere. It's known as the Oort Cloud.

IS THE OORT CLOUD PART OF OUR SOLAR SYSTEM? IT IS HARD TO SAY. THE OBJECTS HERE SIT A LONG, LONG WAY AWAY FROM THE SUN. THIS MEANS THAT THE SUN'S GRAVITATIONAL FORCE ON THEM IS SMALL AND THEY CAN ALSO BE MOVED BY OTHER STARS.

JAN OORT'S IDEA

An astronomer called Jan Oort worked out that millions of icy objects exist in a region way beyond the Kuiper Belt. His theory came from studying "long-period" comets, which take hundreds of years to orbit the Sun. He suggested that most of these start out as frozen objects in a distant region of space – the Oort Cloud.

INSIDE THE CLOUD

Our spacecraft is surrounded by millions of icy objects – some tower over us like mountains. When an object in the Oort Cloud is knocked off orbit by the pull of a star or another body, it hurtles toward the Sun and is transformed into a comet. A comet from the Oort Cloud can take hundreds – even thousands – of years to orbit around the Sun.

IT IS THOUGHT THAT THE BODIES IN THE OORT CLOUD STARTED OFF IN AND AROUND THE PLANETS IN OUR SOLAR SYSTEM. AT SOME POINT, THEY GOT A GRAVITATIONAL KICK OUT TO THEIR CURRENT POSITION.

A MYSTERIOUS ZONE

No one has actually seen the objects in this zone because it is too far from Earth for our current technology to be able to detect clearly – around 15 trillion km or 100,000 AU from the Sun. But scientists believe that the existence of long-period comets provides enough evidence to prove that the Oort Cloud is out there.

VOYAGERS 1 AND 2 ARE THE ONLY SPACECRAFT ANYWHERE NEAR THE OORT CLOUD. IF THEY KEEP TRAVELING AT 60,000 KM PER HOUR, THEY WILL REACH THE START OF THE OORT CLOUD IN AROUND 300 YEARS. BUT IT WILL TAKE 30,000 MORE YEARS TO GET TO THE OTHER SIDE!

1,000–100,000 AU FROM THE SUN

THE SPACE BETWEEN THE STARS

The Oort Cloud is now behind us and we're looking out into deep, dark interstellar space – the region between our star system and other star systems. This point, a very, very long way from Earth, marks the end of our tour of the Solar System.

THE REALLY BRIGHT STAR YOU CAN SEE FAR BEHIND US IS OUR SUN. WE ARE NOW OVER 100,000 ASTRONOMICAL UNITS AWAY FROM IT.

FULL OF GAS

This region is made up of 99% gases – mainly hydrogen and helium – and a teeny 1% of dustlike material. All this matter is spread out so thinly that only around one atom exists per cubic centimeter. It's a different story on Earth – matter at sea level has around 10 BILLION, BILLION atoms per cubic centimeter.

DARK AND COLD

Interstellar space is a long way from the Sun, so it is extremely dark and incredibly cold. The effect of the solar wind and the Sun's magnetic field are no longer felt out here. Interstellar space is controlled by the influences of different stars.

NEBULAE

Interstellar space is the realm of nebulae – huge clouds of dust, gas and red light. A nebula can be created by radiation from a dying star, which excites the gases in interstellar space, causing them to light up. Look very closely as we pass by and you may see small areas where the light is brighter and even emitting other colors. These areas are the stellar nurseries where new stars are born.

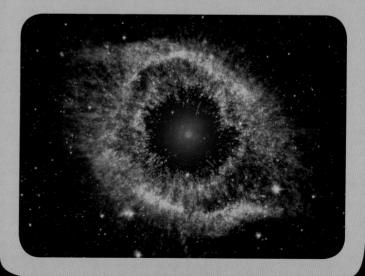

DR. MAGGIE SIGNING OFF

We did it! From the surface of Earth to the edge of the Oort Cloud, this has been an amazing journey and I would like to thank you for making it with me.

But why stop here? Our Sun is one of 300 billion stars in our Milky Way galaxy, and our galaxy is one of around 200 billion galaxies in our universe. There are a LOT of amazing things out there, such as black holes, pulsars and other extraordinary celestial bodies. With the power of thought, we could just keep going and see a whole lot more.

Are you tempted ...?

SHIP'S DATABASE

As well as allowing you to get up close to the wonders of the Solar System, your ship is also equipped with a state-of-the-art database.

You have unlimited access to the database throughout the tour. Here you can find all kinds of fascinating facts, stats and answers, such as how long it would take to travel to each planet and how to understand the vast numbers that we encounter in space. You can also learn a bit more about the pioneering scientists and astronauts who've paved the way for our epic journey.

1 STAR

The Sun is the only star in our Solar System. It is similar to the stars we see in the night sky.

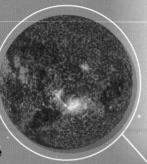

8 PLANETS

5 DWARF PLANETS

For a long time, Pluto was thought to be the ninth planet in the Solar System. Since the discovery of other, similar small worlds, it's been reclassified as a dwarf planet.

COUNTING OUR
SOLAR SYSTEM

There are a surprising number of objects – known as astronomical bodies – which make up our Solar System. At the last count it contained ...

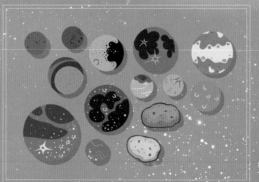

OVER 150 MOONS (AND COUNTING ...)

Scientists are discovering more moons all the time.

Jupiter alone has 53 named moons and another 26 that are waiting to be given official names.

3,564 COMETS

While asteroids are made up of rock and metal, comets are frozen balls of gases and rock. Some have been visited by space probes.

4 MAJOR RING SYSTEMS

The giant planets Saturn, Jupiter, Uranus and Neptune all have ring systems. Some asteroids have them, too.

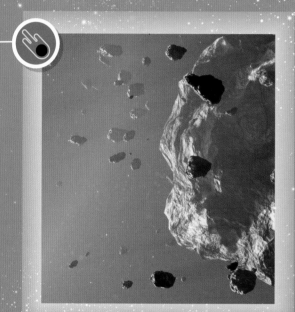

792,353 ASTEROIDS

HOW WE KNOW WHAT'S OUT THERE

The main way to see the universe is to use light. Some telescopes, such as Hubble, collect visible light (that's what we see with our eyes). But other telescopes collect infrared, ultraviolet or radio waves instead – to name a few! These different types of radiation are part of the electromagnetic spectrum.

NUMBERS IN SPACE

The scale of many things out in space is STUPENDOUS, so you need to get used to working with very big numbers.

COUNTING UP

We often count up in 10s, 100s and 1000s. Imagining these amounts is easy – you have TEN fingers, a school may hold HUNDREDS of children and THOUSANDS of people might live in a town.

Now move up the scale and try to picture 10,000s (tens of thousands); 100,000s (hundreds of thousands) and 1,000,000s (millions). Tricky, isn't it?

THESE COMPARISONS MAY HELP …

RICE
1 KG

There are around **50,000** grains in a **1 kilogram** bag of rice.

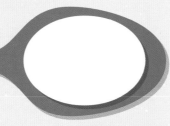

There are around **750,000** crystals in a single teaspoon of sugar – that's three-quarters of a million grains that you can hold in the palm of your hand!

But these numbers are tiny compared to the numbers in space!
OUT THERE WE ENCOUNTER …

10,000,000s
(tens of millions)

100,000,000s
(hundreds of millions)

1,000,000,000s
(billions)

… and the REALLY big one …

1,000,000,000,000s
(trillions).

You also encounter incredibly **ENORMOUS** distances in space. Let's start by imagining a distance close to home.

The Moon is around 384,400 km from Earth.

That distance is about equal to a line of 30 Earths.

Now move up the scale ...

Earth is about 150,000,000 km from the Sun.

Okay, things are starting to get mind-boggling. It's time to introduce your new friend, the AU.

HOORAY FOR ASTRONOMICAL UNITS!

To avoid all those zeros, scientists invented a new unit of measure for distances in space – **THE ASTRONOMICAL UNIT, OR AU.** 1 AU is equal to about 150,000,000 km, the distance from Earth to the Sun.

Let's try another one, this time thinking in AU.
Saturn is about
1,500,000,000 KM
from the Sun
– TEN TIMES the distance from Earth to the Sun.

This means the distance from Saturn to the Sun is about

10 AU
(10 x 150,000,000 km = 1,500,000,000 km).

TRAVEL PLANNER

How long would a real trip around the Solar System take? Here are the journey times from Earth to all the main destinations. These calculations are based on traveling constantly at the top speeds of the fastest vehicles in existence.

FASTEST TRANSPORT	TOP SPEED	Sun 149.6 million km from Earth
SIEMENS VELARO TRAIN	350 km per hour	50 years
BUGATTI VEYRON SPORTS CAR	431 km per hour	40.6 years
BOMBARDIER GLOBAL 6500 JET	1,087 km per hour	16 years
DAWN SPACECRAFT	40,233 km per hour	155 days
JUNO SPACECRAFT	265,000 km per hour	24 days

HOW MANY YEARS?

Use the travel guide to see how many Earth years a trip to these destinations would really take in our vast Solar System. Are we *nearly* there yet?

Uranus 2.7 billion km from Earth	Saturn 1.2 billion km from Earth
428 days	202 days
7.7 years	3.6 years
293 years	138 years
739 years	348 years
910 years	429 years

Neptune 4.3 billion km from Earth	Kuiper Belt 4.3 billion km from Earth
1,452 years	1,450 years
1,179 years	1,177 years
468 years	467 years
12.3 years	12.3 years
1.9 years	1.9 years

TIME, DISTANCE AND SPEED

We measure the time a journey takes by dividing the distance a vehicle must travel by its speed. Spacecraft are designed to travel at incredibly high speeds. They would easily outstrip Earth's fastest vehicles.

Mercury 91 million km from Earth	Venus 41 million km from Earth
30.5 years	14 years
25 years	11 years
10 years	4.5 years
94.5 days	43 days
14 days	6.5 days

Jupiter 628 million km from Earth	Mars 78 million km from Earth
99 days	12 days
1.8 years	81 days
68 years	8 years
170.5 years	21 years
210 years	26 years

	Oort Cloud 15 trillion km from Earth	
1,924 years	34,944 years	4,999,043 years
1,562 years	28,376.5 years	4,059,548 years
619 years	11,251 years	1,609,627 years
16.3 years	296.5 years	42,416 years
2.5 years	45 years	6,439 years

Pluto 5.7 billion km from Earth	Planet Nine 104 billion km from Earth

DR. MAGGIE'S TOP LOCATIONS FOR LIFE IN THE SOLAR SYSTEM

MARS

Scientists have been looking for signs of life on Mars for years. Studies of the surface show that it had liquid water at one time – and some may still exist deep underground.

WHAT IS LIFE?

I'm alive and so are you – but if someone brought us a mysterious specimen to examine, how could we tell if it was alive? A good start would be to see if it had any of the characteristics shared by all living things on Earth:

1. Made of carbon-based matter
2. Needs liquid water
3. Uses energy
4. Grows or reproduces in some way and passes on its genes
5. Responds to stimuli (something that causes growth or activity), such as moving toward sunlight, like plants do

THE GOLDILOCKS ZONE

The Goldilocks Zone is neither too close nor too far from the Sun. Planets here are at just the right temperature for liquid water to exist. Earth sits in the Goldilocks Zone and so does Mars, although its water has now disappeared.

EUROPA, JUPITER'S MOON

The oceans below Europa's icy crust may be capable of hosting simple life forms.

WHAT ARE THE CONDITIONS FOR LIFE?

Living things can only exist in places with the right conditions to support them. The search for life in the Solar System focuses on three key areas:

1. **CARBON**: No problem; this is everywhere in the Solar System.
2. **LIQUID WATER**: This is hard to find because water is only liquid in places with the right temperature and pressure.
3. **ENERGY**: Scientists once thought that energy for life had to come from the Sun. The discovery of simple life forms deep under Earth's oceans showed that energy can come from other sources, too.

ENCELADUS, SATURN'S MOON

In 2005, the space probe Cassini captured photos of frozen geysers erupting from its surface. Scientists think the geysers are powered by warmer liquid water.

PLUTO

The heating actions of gravity between Pluto and its largest moon, Charon, may mean that liquid water – or another liquid chemical that could support life – could exist here. There's no evidence yet, but it's possible!

TITAN, SATURN'S MOON

Cassini's partner probe, Huygens, landed on Titan and found lakes of carbon-based chemicals. In some ways, Titan is like Earth before life existed.

SPACE PEOPLE

Our knowledge of the Solar System today is the result of hundreds of years of study and exploration by scientists, engineers and astronauts from around the world. Here's a shout-out to some pioneering space people you may not have heard about.

CAROLINE HERSCHEL
1750-1848
GERMAN

An astronomer who discovered several comets. Along with her brother, the astronomer William Herschel, she discovered more than 2,400 astronomical objects over a period of 20 years. A crater on the Moon is named after her.

ASAPH HALL
1829-1907
AMERICAN

Hall discovered the moons of Mars – Phobos and Deimos. He also calculated the rotation of Saturn and the mass of Mars.

ANNIE JUMP CANNON
1863-1941
AMERICAN

An astronomer who helped to develop the way we catalog stars.

SUBRAHMANYAN CHANDRASEKHAR
1910-1995   INDIAN-AMERICAN

An astrophysicist who won the 1983 Nobel Prize for Physics for his work on the structure of stars. He proved that black holes had to exist.

RUBY PAYNE-SCOTT

1912-1981   AUSTRALIAN

A physicist who became the first female radio astronomer. She used radio waves to detect solar bursts, and her work led to the discovery of black holes and pulsars, and helped later scientists to understand how solar storms influence weather in space and on Earth.

VALENTINA TERESHKOVA

BORN 1937
RUSSIAN

The first woman to travel into space. She orbited Earth 48 times and remains the only woman to have ever been on a solo space mission.

KATHERINE JOHNSON

BORN 1918
AMERICAN

Johnson is an African-American mathematician whose work at NASA was critical to the success of the Apollo Moon landings.

JOCELYN BELL BURNELL

BORN 1943
BRITISH

An astrophysicist who discovered radio pulsars – rotating stars that send out a beam of light, which can only be seen when the beam is facing Earth. Throughout her career, she's helped and encouraged students who are underrepresented in physics to study the subject she loves.

YVONNE BRILL

1924-2013
CANADIAN-AMERICAN

Brill pioneered the system that keeps communication satellites in orbit – it's still being used to handle phone calls and TV broadcasts around the world. She also helped to develop the rocket engine for the space shuttle.

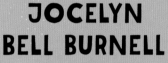

MAE JEMISON

BORN 1956
AMERICAN

An engineer and NASA astronaut. In 1992, she became the first African-American woman to travel into space.

SPACE WORDS

Here are explanations of some handy words we'll encounter on our travels.

asteroid
A lumpy chunk of rock and metal that orbits the Sun.

astronomical unit
A unit of measuring distances in space. One AU is equal to 150 million km, the distance from the Sun to Earth.

atmosphere
A layer of gases that surrounds a planet.

atoms
Tiny particles which make up us and everything that surrounds us.

axis
An imaginary line that an object, such as a planet, turns around.

carbon dioxide (CO_2)
A colorless, odorless molecule made up of the elements carbon and oxygen.

core
The innermost part of a planet.

density
A measurement that compares the amount of mass an object has to its volume.

eclipse
When one object in space blocks the view of another. For instance, during a lunar eclipse, Earth comes between the Sun and the Moon.

electromagnetic spectrum
Electromagnetic radiation is a special kind of radiation that includes visible light, radio waves, gamma rays and X-rays. In all cases, both an electric and magnetic field vary simultaneously. The spectrum is the range of radiation – from shortwave gamma rays at one end to long-wave radio waves at the other.

elliptical
Shaped like an oval (or egg).

equator
An imaginary line around the middle of a planet, which divides it into two equal northern and southern parts.

exoplanet
A planet outside the Solar System.

exosphere
A thin, atmosphere-like volume surrounding a body. On Earth, it sits at the outer edges of the atmosphere.

extravehicular activity
Any activity done by an astronaut outside a spacecraft.

fusion
The process of joining atoms together to produce energy.

galaxy
A group of stars, planets and other bodies held together by gravity.

geyser
A hole in the ground from which jets of water or steam escape.

global warming
The increase in a planet's temperature caused by rising levels of specific gases in the atmosphere.

Goldilocks Zone
The region of our Solar System where temperatures are just right for liquid water to exist.

gravity
A force of attraction between all things with mass.

heliosphere
The region of space surrounding the Sun where the effects of the solar wind are felt. The edge of the heliosphere is called the heliopause.

helium
A colorless, odorless element that is less dense than air. The second most abundant element in the universe.

hemisphere
One half of a sphere. It is usually used to refer to half of the Earth or another spherical body.

hydrogen
The most simple form of an element, it is colorless and odorless. It is the most common element found in the universe.

hydrothermal
Relating to hot water heated beneath a planet's surface.

lander
A vehicle designed to land on another body in space.

liquid water
Water that exists in liquid form.

magnetic field
The volume around a magnetic object in which the magnetic forces due to the object can be felt.

mantle
The layer of a rocky planet or body which lies underneath the crust but around the core.

mass
The measure of how much matter there is in an object.

matter
Any object or substance that takes up space. Solid, liquid, gas and plasma are the four states of matter.

meteor
A small rock that is traveling through the Earth's atmosphere. Before it enters the Earth's atmosphere, it is known as a meteoroid. When it reaches Earth's surface, it is known as a meteorite.

methane
A colorless, odorless molecule made up of hydrogen and carbon. It is sometimes used as a fuel.

microgravity
The apparently very low gravity experienced by objects in orbit around a body, which makes them appear weightless.

molecule
A group of atoms bonded together.

moon
A body that orbits around another planet or other body. Some asteroids have moons.

nebula
A cloud of gases and dust in outer space. Some nebulae are formed when a dying star explodes. Others are regions where new stars form.

orbit
The path of an object around a planet, star or moon.

oxygen
A colorless, odorless element which forms about 20% of the air on Earth and is essential for life.

plasma
Considered to be the fourth state of matter. It is like a gas but contains charged particles. Plasma is the most common state of matter in the universe.

pole
Either the northern or southern end of a body's axis.

probe
A spacecraft, without a human crew, launched into space to collect data that can be sent back to Earth.

protostar
A contracting cloud of gas and dust that represents the early stage in the formation of a star, before fusion has begun.

pulsar
A type of rotating star that emits a beam of radiation which appears to pulse when detected from Earth.

radiation
The emission of energy through space in the form of waves or particles.

satellite
A natural body (such as a moon) or spacecraft that orbits a planet.

solar wind
The flow of charged particles that is released from the Sun and travels throughout the Solar System.

terrestrial planet
Planets with Earthlike characteristics, such as a compact rocky surface. Mercury, Venus, Earth and Mars are terrestrial planets. We think some exoplanets may have these characteristics, too.

universe
The whole of space – all the planets, stars, matter and energy within it.

INDEX

Apollo program, the 5, 117
asteroid belt, the 8, 10, 60-61, 64, 65
asteroids 4, 6, 20, 29, 36, 60-62, 77, 109
astronomical units 9, 61, 86, 89, 101, 111

carbon dioxide 47, 49, 53
Cassini-Huygens 65, 74, 80-81, 115
comets 6, 20, 36, 62, 77, 100-101, 109, 116
craters 29, 38, 39, 40, 42, 73, 89, 94, 95
Curiosity 51, 58

Dawn 61, 112
Deep Space Network, the 65
dwarf planets 6, 8, 61, 90-95, 108
   Eris 90, 93
   Ceres 61
   Haumea 90-91, 93
   Makemake 90, 93

Earth 4, 6, 8, 11, 12, 13, 14, 15, 16, 18-21, 22, 24, 25, 27, 28, 29, 30, 31, 32, 33, 35, 38, 39, 40, 41, 44, 45, 46, 47, 49, 53, 54, 55, 56, 57, 58, 59, 61, 63, 65, 66, 67, 68, 69, 70, 72, 75, 77, 78, 80, 81, 83, 85, 86-87, 88, 92, 95, 96, 99, 101, 102, 105, 111, 113, 114, 117
Einstein, Albert 5, 37
ExoMars 58
exoplanets 7
exosphere 40, 73

fusion 37

Galilei, Galileo 72
global warming 49
Goldilocks Zone, the 20, 114
gravity 4, 6, 12-13, 14-15, 22, 27, 29, 34, 36-37, 66, 68, 77, 83, 85, 95, 99
Great Red Spot, the 68, 70-71

heliopause 6, 9, 98-99
heliosphere 98

International Space Station, the 14-15, 20, 22-23, 24, 25
interstellar space 9, 64, 102-103

Juno 71, 112
Jupiter 6, 8, 11, 20, 60, 61, 64, 66-73, 74, 76, 77, 85, 113, 115
   moons of:
   Europa 72-73, 115
   Io 72
   Callisto 73
   Ganymede 73

Kuiper Belt, the 8, 11, 61, 78, 83, 90-91, 93, 96-97, 99, 100, 112

light curves 91
liquid water 20, 27, 58, 72, 81, 114, 115

Maat Mons 47
Mariner 10 50
Mariner 2 50
Mariner 4 7, 51
Mars 6, 7, 8, 10, 40, 51, 52-59, 61, 96, 113, 114, 116
   moons of:
   Deimos 54, 116
   Phobos 54, 116
Mercury 6, 8, 10, 38-43, 48, 49, 73, 78, 92, 93, 133
meteorites 4, 21, 29, 63
meteoroids 63
meteors 40, 63
microgravity 15
Milky Way, the 7, 35, 105
Moon, the 4, 5, 8, 11, 14, 21, 24, 26-31, 32, 33, 39, 40, 48-49, 72, 78, 85, 116
moonlets 78-79
moons 4, 6, 14, 32, 54, 62, 65, 72-73, 74, 77, 78-79, 80-81, 82, 83, 85, 86, 88-89, 95, 108
moonwalking 8, 28
Mount Everest 47, 56

NASA 50, 51, 52, 58, 65, 117
nebulae 36-37, 103
Neptune 6, 8, 11, 64, 76, 86-89, 93, 96, 112
   moons of:
   Triton 86, 88, 89
New Horizons 11, 90

Olympus Mons 53, 56-57
Oort Cloud, the 5, 9, 11, 100-101, 102, 105, 113
orbits 6, 13, 14-15, 22, 24, 30, 31, 34, 35, 38-39, 43, 55, 60, 61, 62, 65, 75, 77, 78, 80, 86, 90, 96-97, 99, 100, 101
oxygen 19, 21, 53, 57, 73

Philae 65
Pioneer 10 64
Pioneer 11 64
Planet Nine 9, 96-97, 113
Pluto 8, 9, 11, 90, 92-95, 113, 115
   moons of:
   Charon 95, 115
protostars 37

radiation 21, 32, 37, 58, 103, 109
ring systems 8, 68, 74-77, 79, 82-83, 109
ROSCOSMOS 50
Rosetta 65

satellites 4, 14-15, 16, 24, 25, 27, 33, 117
Saturn 6, 8, 10, 64, 65, 74-81, 111, 112, 115
   moons of:
   Atlas 79
   Dione 78
   Enceladus 74, 81, 115
   Hyperion 79
   Iapetus 78
   Pan 79
   Phoebe 78
   Titan 65, 78, 80-81, 115
shooting stars 25, 63
Sojourner 51
solar flares 32, 34, 99
Solar System, the 4-5, 6-7, 8-9, 11, 19, 20, 32, 33, 34, 35, 36, 39, 40, 43, 45, 47, 48, 53, 55, 56, 61, 62, 64, 65, 66, 67, 69, 72, 74, 75, 76, 78, 80, 81, 83, 84, 85, 86, 88, 89, 90, 91, 93, 96, 97, 98-99, 100-101, 102, 116
solar wind 21, 62, 98-99, 103
Soyuz 23
Space Race 4
stars 6, 7, 35, 41, 47, 89, 97, 98, 100, 102, 103, 105, 108, 116
stellar nurseries 36, 103
storms 34, 54-55, 58, 66, 70-71, 77, 85, 88
Sun, the 6, 8, 11, 20, 28, 32-37, 39, 41, 42, 43, 47, 48, 49, 61, 62, 63, 74, 86, 95, 96-97, 100-101, 102-103, 105, 111, 112, 114, 115
sunspots 34

terrestrial planets 40
tholins 95

Ultima Thule 90
Uranus 6, 8, 10, 64, 76, 82-85, 89, 112
   moons of:
   Miranda 83, 85

Valles Marineris 55
Venera program 50
   Venera 3 46
   Venera 7 46
Venus 6, 8, 10, 40, 44-49, 50, 53, 113
Verona Rupes 85
Viking 1 51
Viking 2 51
volcanoes 47, 53, 56-57, 72
Voyager 1 64, 101
Voyager 2 64, 83, 88, 101

water ice 42, 77, 95

DEDICATION

To my wonderful, talented daughter, Lori, and all the other children who have been the inspiration for this book. I thank you all and hope that we grown-ups can see and appreciate the world and beyond as you do.

To my husband, family and friends, and all those at Buster Books – thank you for all of your help and support.

To Chelen Écija for creating such vivid images that made me gasp and realize my visions of what is out there.

To my fellow space travelers – through the thought experiment in this book we can dream of space, but may we one day get out there for real.

Happy travels.